SCIENCE
IN THE
SOUL

SELECTED WRITINGS OF
A PASSIONATE RATIONALIST

RICHARD
DAWKINS

EDITED BY GILLIAN SOMERSCALES

BLACK SWAN

TRANSWORLD PUBLISHERS
61–63 Uxbridge Road, London W5 5SA
www.penguin.co.uk

Transworld is part of the Penguin Random House group of companies
whose addresses can be found at global.penguinrandomhouse.com

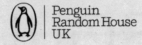

First published in Great Britain in 2017 by Bantam Press
an imprint of Transworld Publishers
Black Swan edition published 2018

A CIP catalogue record for this book
is available from the British Library.

ISBN 9781784162016

Typeset in 10.32/12.9pt Minon Pro by Jouve (UK), Milton Keynes.
Printed and bound in Great Britain by Clays Ltd, Bungay, Suffolk.

Penguin Random House is committed to a sustainable future
for our business, our readers and our planet. This book is made
from Forest Stewardship Council® certified paper.

13 5 7 9 10 8 6 4 2

In memory of
Christopher Hitchens

Contents

V Living in the real world

VI The sacred truth of nature

VII Laughing at live dragons

VIII No man is an island

Author's introduction

I AM WRITING THIS two days after a breathtaking visit to Arizona's Grand Canyon ('breathtaking' still hasn't gone the way of 'awesome' although I fear it may). To many Native American tribes the Grand Canyon is a sacred place: site of numerous origin myths from the Havasupai to the Zuni; hushed repose of the Hopi dead. If I were forced to choose a religion, that's the kind of religion I could go for. The Grand Canyon confers stature on a religion, outclassing the petty smallness of the Abrahamics, the three squabbling cults which, through historical accident, still afflict the world.

In the dark night I walked out along the south rim of the canyon, lay down on a low wall and gazed up at the Milky Way. I was looking back in time, witnessing a scene from a hundred thousand years ago – for that is when the light set out on its long quest to dive through my pupils and spark my retinas. At dawn the following morning I returned to the spot, shuddered with vertigo as I realized where I had been lying in the dark, and looked down towards the canyon's floor. Again I was gazing into the past, two billion years in this case, back to a time when only microbes stirred sightless beneath the Milky Way. If Hopi souls were sleeping in that majestic hush they were joined by the rockbound ghosts of trilobites and crinoids, brachiopods and belemnites, ammonites, even dinosaurs.

Was there some point in the mile-long evolutionary progression up the canyon's strata when something you could call a 'soul' sprang into existence, like a light suddenly switched on? Or did 'the soul' creep stealthily into the world: a dim thousandth of a soul in a pulsating tube-worm, a tenth of a soul in a coelacanth, half a soul in

1

a tarsier, then a typical human soul, eventually a soul on the scale of a Beethoven or a Mandela? Or is it just silly to speak of souls at all?

Not silly if you mean something like an overwhelming sense of subjective, personal identity. Each one of us knows we possess it even if, as many modern thinkers aver, it is an illusion – an illusion constructed, as Darwinians might speculate, because a coherent agency of singular purpose helps us to survive.

Visual illusions such as the Necker Cube—

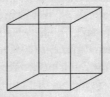

—or the Penrose Impossible Triangle—

—or the Hollow Mask illusion demonstrate that the 'reality' we see consists of constrained models constructed in the brain. The Necker Cube's two-dimensional pattern of lines on paper is compatible with two alternative constructions of a three-dimensional cube, and the brain adopts the two models in turn: the alternation is palpable and its frequency can even be measured. The Penrose Triangle's lines on paper are incompatible with any real-world object. These illusions tease the brain's model-construction software, thereby revealing its existence.

In the same way, the brain constructs in software the useful illusion of personal identity, an 'I' apparently residing just behind the eyes, an 'agent' taking decisions by free will, a unitary personality, seeking goals and feeling emotions. The construction of personhood takes place progressively in early childhood, perhaps by the joining up of previously disparate fragments. Some psychological disorders are interpreted as 'split personality', failure of the fragments to unite. It's a not unreasonable speculation that the progressive growth of consciousness in the infant mirrors a similar progression over the longer timescale of evolution. Does a fish, say, have a rudimentary feeling of conscious personhood, on something like the same level as a human baby?

We can speculate on the evolution of the soul, but only if we use the word to mean something like the constructed internal model of a 'self'. Things are very different if, by 'soul', we mean a spook that survives bodily death. Personal identity is an emergent consequence of material brain activity and it must disintegrate, eventually reverting to its pre-birth nothingness, when the brain decays. But there are poetic usages of 'soul' and related words that I am unashamed to embrace. In an essay published in my earlier anthology *A Devil's Chaplain*, I deployed such words to extol a great teacher, F. W. Sanderson, headmaster of my old school before I was born. Notwithstanding the ever-present risk of misunderstanding, I wrote of the 'spirit' and the 'ghost' of the dead Sanderson:

> His spirit lived on at Oundle. His immediate successor, Kenneth Fisher, was chairing a staff meeting when there was a timid knock on the door and a small boy came in: 'Please, sir, there are Black Terns down by the river.' 'This can wait,' said Fisher decisively to the assembled committee. He rose from the Chair, seized his binoculars from the door and cycled off in the company of the small ornithologist, and – one can't help imagining – with the benign, ruddy-faced ghost of Sanderson beaming in their wake.

I went on to refer to the 'shade' of Sanderson, after describing another scene, from my own education, when an inspiring science teacher, Ioan Thomas (who came to the school because he admired Sanderson although he was too young to have met him) dramatically taught us the value of admitting ignorance. He asked us, one by one, a question to which we all wildly guessed answers. Finally, our curiosity aroused, we clamoured ('Sir! Sir!') for the true answer. Mr Thomas waited dramatically for silence and then spoke slowly and distinctly, pausing for effect between each word. 'I don't know! I . . . don't . . . know!'

> Again the fatherly shade of Sanderson chuckled in the corner, and none of us will have forgotten that lesson. What matters is not the facts but how you discover and think about them: education in the true sense, very different from today's assessment-mad exam culture.

Was there a risk that readers of my earlier essay might misunderstand that 'spirit' of Sanderson 'living on'; his benign, ruddy-faced 'ghost' beaming; or his 'shade' chuckling in the corner? I don't think so, although, God knows (there we go again), there's enough eager appetite for misunderstanding out there.

I have to recognize that the same risk, born of the same eagerness, stalks this volume's title. *Science in the Soul*. What does it mean?

Let me sidestep before attempting to answer. I think it's high time the Nobel Prize for Literature was awarded to a scientist. I'm sorry to say the nearest precedent is a very poor example: Henri Bergson, more of a mystic than a true scientist, whose vitalistic *élan vital* was lampooned in Julian Huxley's satirical railway train propelled by *élan locomotif*. But seriously, why not a real scientist for the Literature prize? Although, alas, he is no longer with us to receive it, who would deny that Carl Sagan's writing is of Nobel literary quality, up there with the great novelists, historians and poets? How about Loren Eiseley? Lewis Thomas? Peter Medawar? Stephen Jay Gould? Jacob Bronowski? D'Arcy Thompson?

Whatever the merits of individual authors whom we might name, isn't science itself a worthy subject for the best writers, more than capable of inspiring great literature? And whatever the qualities are that make science so – those same qualities that make for great poetry and Nobel-winning novels – don't we have here a good approach to the meaning of 'soul'?

'Spiritual' is another word that could be used to describe Saganesque literary science. It is widely thought that physicists are more likely to self-identify as religious than biologists. There's even statistical evidence of this from the Fellows of both the Royal Society of London and the US National Academy of Sciences. But experience suggests that if you probe these elite scientists further you'll find that even the 10 per cent who profess some kind of religiosity have, in most cases, no supernatural beliefs, no god, no creator, no aspiration to an afterlife. What they do have – and they'll say so if pressed – is a 'spiritual' awareness. They may be fond of the hackneyed phrase 'awe and wonder', and who can blame them? They may quote, as I do in these pages, the Indian astrophysicist Subrahmanyan Chandrasekhar 'shuddering before the beautiful', or the American physicist John Archibald Wheeler:

> Behind it all is surely an idea so simple, so beautiful, that when we grasp it – in a decade, a century, or a millennium – we will all say to each other, how could it have been otherwise? How could we have been so blind?

Einstein himself made it very clear that, though spiritual, he believed in no kind of personal god.

> It was, of course, a lie what you read about my religious convictions, a lie which is being systematically repeated. I do not believe in a personal God and I have never denied this but have expressed it clearly. If something is in me which can be called religious then it is the unbounded admiration for the structure of the world so far as our science can reveal it.

And on another occasion:

> I am a deeply religious nonbeliever – this is a somewhat new kind of religion.

Though I wouldn't use the exact phrase, it is in this sense of a 'deeply religious nonbeliever' that I consider myself a 'spiritual' person, and it is in this sense that I unapologetically use 'soul' in the title of this book.

Science is both wonderful and necessary. Wonderful for the soul – in contemplation, say, of deep space and deep time from the rim of the Grand Canyon. But also necessary: for society, for our well-being, for our short-term and long-term future. And both aspects are represented in this anthology.

I've been a science educator all my adult life, and most of the essays collected here stem from the years when I was the inaugural Charles Simonyi Professor of the Public Understanding of Science. When promoting science, I've long been an advocate of what I call the Carl Sagan school of thought: the visionary, poetic side of science, science to stir the imagination, as opposed to the 'non-stick frying pan' school of thought. By the latter I mean the tendency to justify the expense of, for example, space exploration by reference to spin-offs such as the non-stick frying pan – a tendency I have compared to an attempt to justify music as good exercise for the violinist's right arm. It's cheap and demeaning, and I suppose my satirical description could be accused of exaggerating the cheapness. But I still use it to express my preference for the romance of science. To justify space exploration I would rather invoke what Arthur C. Clarke extolled and John Wyndham named as 'the outward urge', the modern version of the urge that drove Magellan, Columbus and Vasco da Gama to exploring the unknown. But yes, 'non-stick frying pan' is unfairly demeaning to the school of thought that it labels in my shorthand; and it is to the serious, practical value of science in our society that I now turn, for that is what many of the essays in this book are about. Science really matters for life – and by 'science' I mean not just scientific facts, but the scientific way of thinking.

I write this in November 2016, bleak month in a bleak year when the phrase 'barbarians at the gates' tempts without irony. Within the gates, rather, for the calamities that have struck the two most populous nations in the English speaking world in 2016 are self-inflicted: wounds dealt not by an earthquake or a military *coup d'état* but by the democratic process itself. More than ever, reason needs to take centre stage.

Far be it from me to devalue emotion – I love music, literature, poetry and the warmth, both mental and physical, of human affection – but emotion should know its place. Political decisions, decisions of state, policies for the future, should flow from clear-thinking, rational consideration of all the options, the evidence bearing upon them, and their likely consequences. Gut feelings, even when they don't arise from the stirred dark waters of xenophobia, misogyny or other blind prejudice, should stay out of the voting booth. For some time now, such murky emotions have remained largely under the surface. But in 2016 political campaigns on both sides of the Atlantic brought them into the open, made them, if not respectable, at least freely expressed. Demagogues led by example and proclaimed open season for prejudices which had for half a century been shamed into hole-in-corner secrecy.

Whatever may be the innermost feelings of individual scientists, science itself works by rigorous adherence to objective values. There is objective truth out there and it is our business to find it. Science has in place disciplined precautions against personal bias, confirmation bias, prejudgement of issues before the facts are in. Experiments are repeated, double-blind trials exclude the pardonable desires of scientists to be proved right – and the more laudable bending over backwards to maximize our opportunity to be proved wrong. An experiment done in New York can be replicated by a lab in New Delhi and we expect the conclusion to be the same regardless of geography or the cultural or historic biases of the scientists. Would that other academic disciplines such as theology could say the same. Philosophers happily speak of 'continental philosophy' as opposed to 'analytical philosophy'. Philosophy departments in American or British universities might seek a new appointment to 'cover the continental tradition'. Can you imagine a science department advertising for a new professor to cover 'continental chemistry'? Or 'the Eastern tradition in biology'? The very idea is a bad joke. That says something about the values of science and is not kind to those of philosophy.

Starting with the romance of science and the 'outward urge', then, I have moved on to the values of science and the scientific way of thinking. Some might think it strange to leave the practical usefulness of scientific knowledge till last, but that ordering does reflect my personal priorities. Certainly such medical boons as vaccination, antibiotics and anaesthetics are hugely important, and they are too well known to need rehearsing here. The same goes for climate change (dire warnings there may already be too late) and the Darwinian evolution of antibiotic resistance. But I will pick out for attention here one further warning, less immediate and less well known. It neatly joins the three themes of outward urge, scientific usefulness and the scientific way of thinking. I refer to the inevitable, though not necessarily imminent, danger of a catastrophic collision with a large extraterrestrial object, most likely displaced from the asteroid belt by the gravitational influence of Jupiter.

The dinosaurs, with the notable exception of birds, were wiped out by a massive bolt from space, of a kind which, sooner or later, will strike again. The circumstantial evidence is now strong that a huge meteorite or comet struck the Yucatán peninsula some sixty-six million years ago. The mass of this object (as large as a substantial mountain) and its velocity (perhaps 40,000 miles per hour) would on impact have generated energy equivalent, according to plausible estimates, to several billion Hiroshima bombs exploding together. The scorching temperature and prodigious blast of that initial impact would have been followed by a prolonged 'nuclear winter', lasting perhaps a decade. Together these events killed all the non-bird dinosaurs, plus pterosaurs, ichthyosaurs, plesiosaurs, ammonites, most fish and many other creatures. Fortunately for us a few mammals survived, perhaps protected because they were hibernating in their equivalent of underground bunkers.

A catastrophe on the same scale will threaten again. Nobody knows when, for they strike at random. There is no sense in which they get more likely as the interval between them gets longer. It could happen in our lifetime, but that's unlikely because the average interval between such mega-impacts is of the order of a hundred million years. Smaller – but still dangerous – asteroids, large enough to destroy a city like Hiroshima, hit the Earth about once every century or two. The reason we don't fret about them is that most of our planet's surface is uninhabited. And again, of course, they don't strike regularly such that you can look at the calendar and say: 'We're about due for another one.'

For advice and information on these matters I am indebted to the famous astronaut Rusty Schweickart, who has become the most high-profile advocate of taking the risk seriously and trying to do something about it. What can we do about it? What could the dinosaurs have done if they'd had telescopes, engineers and mathematicians?

The first task is to detect an incoming projectile. 'Incoming' gives a misleading impression of the nature of the problem. These are not speeding bullets heading straight towards us and looming up as they approach. The Earth and the projectile are both in elliptical orbits around the sun. Having detected an asteroid, we need to measure its orbit – which we can do with increasing accuracy the more readings we take into account – and calculate whether at some date, perhaps decades ahead, a future cycle of the asteroid's orbit will coincide with a future cycle of our own. Once an asteroid is detected and its orbit accurately plotted, the rest is mathematics.

The moon's pockmarked face presents a disquieting image of the ravages we are spared because of the Earth's protective atmosphere. The statistical distribution of moon craters of various diameters gives us a reading of what's out there, a baseline against which to compare our meagre success in spotting projectiles ahead of time.

The larger the asteroid, the easier it is to detect. Since small ones – including city-destroying 'small' ones – are hard to detect in the first place, it is entirely possible that we might get very little warning, or none at all. We need to improve our ability to detect asteroids. And that means increasing the number of wide-field watchdog telescopes looking for them, including infra-red telescopes in orbit beyond the reach of distortion caused by the Earth's atmosphere.

Having identified a dangerous asteroid whose orbit threatens eventually to intersect with ours, what do we do then? We need to change its orbit, either by speeding it up so it goes into a larger orbit and therefore arrives at the rendezvous too late to collide, or slowing it down so its orbit contracts and it arrives too early. Surprisingly, a very small change in velocity will suffice, in either direction: as little as 0.025 miles per hour. Without even resorting to high explosives, this can be achieved using existing – though expensive – technologies, technologies not unrelated to the spectacular achievement of the European Space Agency's Rosetta mission to land a spacecraft on a comet, twelve years after its launch in 2004. You see what I meant when I spoke of uniting the 'outward urge' of the imagination with the sober practicalities of useful science and the rigour of the scientific way of thinking? And this detailed example illustrates another feature of the scientific way of thinking, another virtue of what we might call the soul of science. Who but a scientist would accurately predict the exact moment of a worldwide catastrophe a hundred thousand years in the future and formulate a high-precision plan to prevent it?

Despite the timespan over which these essays were written, I find little that I would change today. I could have deleted all reference to the dates of original publication but I chose not to. A few of these pieces are speeches made on particular occasions, for example in opening an exhibition or eulogizing a dead person. I have left them untouched, as they were originally spoken. They retain their intrinsic immediacy, which would be lost were I to edit out all contemporary allusions. I have confined updatings to footnotes and afterwords – brief additions and reflections that could perhaps be read alongside the main texts as a dialogue between me today and the author of the original article. To facilitate such a reading, the footnotes have been set in larger type than is customary for academic footnotes or endnotes.

Gillian Somerscales and I have selected forty-one of my essays, speeches and journalistic writings and grouped them into eight sections. In addition to science itself, they include my reflections on the values of science, the history of science and the role of science in society; some polemics, a little gentle crystal-ball gazing, some satire and humour, and some personal sadnesses which I hope stop short of self-indulgence. Each section begins with Gillian's own sensitive introduction. For me to add to these would be superfluous but I have, as explained above, added my own footnotes and afterwords.

When we were debating various titles for this book, *Science in* (or *for*) *the Soul* was the front-runner, tentatively favoured by both Gillian and me over a large field of competitors. I'm not known for my faith in omens, but I have to admit that what finally swung me was the rediscovery, while cataloguing my library in August 2016, of a delightful little book by Michael Shermer. Called *The Soul of Science*, it was dedicated 'To Richard Dawkins, for giving science its soul'. The serendipity was almost as great as the pleasure, and neither Gillian nor I had any further doubts as to what we should call this book.

My gratitude to Gillian herself is unbounded. In addition I would like to thank Susanna Wadeson of Transworld and Hilary Redmon of Penguin Random House USA for their enthusiastic belief in the project and for their helpful suggestions. Miranda Hale's internet expertise helped Gillian track down forgotten essays. It's in the nature of an anthology whose entries span many years that debts of gratitude span the same years. They were acknowledged in the original articles. I hope it will be understood that I cannot repeat them all here. The same applies to bibliographic citations. Readers interested in following them can look them up in the original articles, full details of which are given in the list at the back of the book.

Editor's introduction

RICHARD DAWKINS has always defied categorization. One eminent biologist of mathematical bent reviewing *The Selfish Gene* and *The Extended Phenotype* was startled to find scientific work apparently free of logical errors and yet containing not a single line of mathematics; he could come to no other conclusion than that, incomprehensible as it seemed to him, 'Dawkins . . . apparently thinks in prose'.

Thank goodness he does. For had he not thought in prose – taught in prose, reflected in prose, wondered in prose, argued in prose – we should not have the exhilarating range of work produced by this most versatile of scientist communicators. Not just his thirteen books, whose qualities I need not rehearse here, but the *embarras de richesses* of shorter writing on many different platforms – daily newspapers and scientific journals, lecture halls and online salons, periodicals and polemics, reviews and retrospectives – from which he and I have distilled this collection. It includes alongside much recent work some vintage earlier pieces, mining the rich seams laid down both before and after publication of his first anthology, *A Devil's Chaplain*.

Given his reputation as a controversialist, it seems to me all the more important to pay due attention to Richard Dawkins' work as a maker of connections, tirelessly throwing word-bridges across the chasm between scientific discourse and the broadest range of public debates. I see him as an egalitarian elitist, dedicated to making complex science not just accessible but *intelligible* – and without 'dumbing down', with a constant insistence on clarity and accuracy, using language as a precision tool, a surgical instrument.

If he also uses language as a rapier, and sometimes a bludgeon, it is to puncture obfuscation and pretension, sweep distraction and muddle out of the way. He has a horror of the fake, whether it be false belief, science, politics or emotion. As I read and reread the candidate pieces for inclusion in this volume, I conceived of a group I called the 'darts': short, pointed pieces, sometimes funny, sometimes blazingly angry, sometimes heartbreakingly poignant or breathtakingly impolite. I was tempted to present a selection of these as a group of their own, but on reflection chose instead to situate a few of them among the longer, more reflective and sustained essays, both better to convey the range of writing overall and to offer the reader immediate experience of the changes of pace and tone that epitomize the buzz of reading Dawkins.

There are extremities here of delight and derision; anger, too – but never at what is said against himself, always at the harm done to others: especially children, non-human animals and people oppressed for contravening the dictates of authority. That anger, and the sadness that drifts behind it for all that is damaged and lost, are for me reminders – and I stress the perception is mine, not Richard's – of the tragic aspect of his writing and speaking life since *The Selfish Gene*. If 'tragic' seems too strong a word, consider this. In that first explosive book he explained how evolution by natural selection proceeds through a logic that expresses itself in relentlessly self-seeking behaviour on the part of the tiny replicators by which living beings are constructed. He then pointed out that humans alone have the power to overcome the dictates of our selfish replicating molecules, to take ourselves and the world in hand, to conceive of the future and then influence it. We are the first species able to be *un*selfish. That's some clarion call. And here's the tragedy: instead of being able thereafter to devote his manifold talents to exhorting humanity to use the precious attribute of consciousness and the ever-increasing insights of science and reason to rise above the selfish drives of our evolutionary programming, he's had to divert much of that energy and skill to persuading people to accept the truth of evolution at all. A grim job, perhaps, but someone's got to do it: for, as he says, 'nature can't sue'. And, as he remarks in one of the pieces reproduced here, 'I have . . . learned that rigorous commonsense is by no means obvious to much of the world. Indeed, commonsense sometimes requires ceaseless vigilance in its defence.' Richard Dawkins is not only reason's prophet; he is our ceaseless watchman.

It's a shame that so many of the adjectives associated with rigour and clarity – remorseless, ruthless, merciless – are so brutal, when Richard's principles are shot through and through with compassion, generosity, kindness. Even his criticism, stringent in judgement, is also astringent in wit – as when he refers in a letter to the Prime Minister to 'Baroness Warsi, your Minister without Portfolio (and without election)', or ventriloquizes a Blair acolyte promoting his boss's promotion of religious diversity: 'We shall support the introduction of Sharia courts, but on a strictly voluntary basis – only for those whose husbands and fathers freely choose it.'

I prefer to use images of clarity: incisiveness, forensic attention to logic and detail, piercing illumination. And I prefer to call his writing athletic rather than muscular – an instrument not just of force and strength but of flexibility, adaptable to pretty much any audience, reader or topic. There aren't many writers, indeed, who manage to combine power and subtlety, impact and exactitude, with such elegance and humour.

I first worked with Richard Dawkins on *The God Delusion*, over a decade ago. If readers of what follows here come to appreciate not only the writer's clarity of thought and facility of expression, the fearlessness with which he confronts very large elephants in very small rooms, the energy with which he devotes himself to explication of the complex and the beautiful in science, but also some of the generosity, kindness and courtesy that have characterized all my dealings with Richard over the years since that first collaboration, then the present volume will have achieved one of its aims.

It will have achieved another if it embodies a condition felicitously described in one of the essays reproduced here, where 'harmonious parts flourish in the presence of each other, and the illusion of a harmonious whole emerges'. Indeed, it is my belief that the harmony resounding from this collection is no illusion, but the echo of one of the most vibrant, and vital, voices of our times.

G.S.

I
THE VALUE(S) OF SCIENCE

WE BEGIN AT THE HEART OF THE MATTER, with science: what it is, what it does, how it is (best) done. Richard's 1997 Amnesty lecture, 'The values of science and the science of values', is a wonderful portmanteau piece, covering a huge amount of ground and trailing several themes developed elsewhere in this collection: the overriding respect of science for objective truth; the moral weight attached to the capacity to suffer, and the dangers of 'speciesism'; a telling emphasis on key distinctions, as between 'using rhetoric to bring out what you believe is really there, and using rhetoric knowingly to cover up what is really there'. This is the voice of the scientist communicator, the determined believer in marshalling language to *convey* truth, not to create an artificial 'truth'. The very first paragraph makes a careful distinction: the values that underpin science are one thing, a proud and precious set of principles to be defended, for on them depends the perpetuation of our civilization; the attempt to derive values *from* scientific knowledge is an altogether different and more suspect enterprise. We must have the courage to admit that we start in an ethical vacuum; that we invent our own values.

The writer of this lecture is no fact-bound Gradgrind, no dry bean (or bone) counter. The passages on the aesthetic value of science, the poetic vision of Carl Sagan, Subrahmanyan Chandrasekhar's 'shuddering before the beautiful' – these epitomize passionate celebration of the glories, beauties and potentialities of science to bring joy to our lives and hope to our futures.

We then make a change of pace and platform as the register shifts from the extended and reflective to the pithy and pointed:

what I like to think of as the Dawkins Dart. Here, with steely courtesy, Richard pursues several points made in his Amnesty lecture in admonishing Britain's next monarch on the perils of following the lead of 'inner wisdom' rather than evidence-based science. Typically, he does not absolve humans from using their judgement in respect of the possibilities offered by science and technology: 'one worrying aspect of the hysterical opposition to the *possible* risks from GM crops is that it diverts attention from *definite* dangers which are already well understood but largely ignored'.

The third piece in this section, 'Science and sensibility', is another wide-ranging lecture, delivered with a characteristic combination of gravitas and sparkle. Here too we see the messianic enthusiasm for science – tempered by a sober reflection on how far we *could* have come by the millennium, and the distances we have not covered. Typically, this is conceived as a recipe not for despair but for redoubled effort.

And where did all this unquenchable curiosity, this hunger for knowledge, this campaigning compassion come from? The section closes with 'Dolittle and Darwin', an affectionate look back at some of the influences that fed into a child's education in the values of science – including a lesson in distinguishing core values from their temporary historical and cultural coloration.

Through all these disparate pieces, the key messages reverberate clearly. It's no good shooting the messenger, no good turning to illusory comforts, no good confusing *is* with *ought* or with what you *might like to be the case*. They are ultimately positive messages: a clear, sustained focus on how things work, coupled with the intelligent imagination of the incurably curious, will yield insights that inform, challenge and stimulate. And so science continues to develop, understanding to grow, knowledge to expand. Taken together, these pieces offer a manifesto for science and a call to arms in its cause.

G.S.

The values of science and the science of values*

THE VALUES OF SCIENCE; what does this mean? In a weak sense I shall mean – and shall take a sympathetic view of – the values that scientists might be expected to hold, insofar as these are influenced by their profession. There is also a strong meaning, in which scientific knowledge is used directly to derive values as if from a holy book. Values in this sense I shall strongly[†] repudiate. The book of nature may be no worse than a traditional holy book as a source of values to live by, but that isn't saying much.

* The Oxford Amnesty Lectures are an annual series, given in the Sheldonian Theatre in aid of Amnesty International. Each year the lectures are collected in a book, edited by an Oxford academic. In 1997 the convenor and editor was Wes Williams and the chosen theme was 'the values of science'. Among the lecturers were Daniel Dennett, Nicholas Humphrey, George Monbiot and Jonathan Rée. Mine was the second of the series of seven and the text is reproduced here.

† Had Sam Harris's thought-provoking book *The Moral Landscape* been published at the time of this lecture, I would have deleted 'strongly'. Harris makes a persuasive case that there are some actions, for example infliction of acute suffering, which it would be perverse to deny are immoral, and that science can play a crucial role in identifying them. A worthwhile case can be made that the fact–value distinction has been oversold. (For full publication details of books referred to in the text and notes, please see the bibliography at the back of this volume.)

The science of values – the other half of my title – means the scientific study of where our values come from. This in itself should be value-free, an academic question, not obviously more contentious than the question of where our bones come from. The conclusion might be that our values owe nothing to our evolutionary history, but that is not the conclusion I shall reach.

The values of science in the weak sense

I doubt that scientists in private are less (or more) likely to cheat their spouses or their tax inspectors than anybody else. But in their professional life scientists do have special reasons for valuing simple truth. The profession is founded on a belief that there is such a thing as objective truth which transcends cultural variety, and that if two scientists ask the same question they should converge upon the same truth regardless of their prior beliefs or cultural background or even, within limits, ability. This is not contradicted by the widely rehearsed philosophical belief that scientists don't prove truths but advance hypotheses which they fail to *dis*prove. The philosopher may persuade us that our facts are only undisproved theories, but there are some theories we shall bet our shirt on never being disproved, and these are what we ordinarily call true.* Different scientists, widely separated geographically and culturally, will tend to converge upon the same undisproved theories.

This view of the world is poles away from fashionable prattlings like the following:

> There's no such thing as objective truth. We make our own truth.
> There's no such thing as objective reality. We make our own reality.

---

* I like Steve Gould's way of putting it. 'In science, "fact" can only mean "confirmed to such a degree that it would be perverse to withhold provisional assent." I suppose that apples might start to rise tomorrow, but the possibility does not merit equal time in physics classrooms' ('Evolution as fact and theory', in *Hen's Teeth and Horse's Toes*).

There are spiritual, mystical, or inner ways of knowing that are superior to our ordinary ways of knowing.* If an experience seems real, it is real. If an idea feels right to you, it is right. We are incapable of acquiring knowledge of the true nature of reality. Science itself is irrational or mystical. It's just another faith or belief system or myth, with no more justification than any other. It doesn't matter whether beliefs are true or not, as long as they're meaningful to you.†

That way madness lies. I can best exemplify the values of one scientist by saying that, if there comes a time when everybody thinks like that, I shall not wish to go on living. We shall have entered a new Dark Age, albeit not one 'made more sinister and more protracted by the lights of perverted science'‡ – because there won't be any science to pervert.

Yes, Newton's Law of Gravitation is only an approximation, and maybe Einstein's General Theory will in due season be superseded. But this does not lower them into the same league as medieval witchcraft or tribal superstition. Newton's laws are approximations that you can stake your life on, and we regularly do. When it comes to flying, does your cultural relativist bet his life on levitation or physics, Magic Carpet or McDonnell Douglas? It doesn't matter

---

* Professors of 'Women's Studies' are occasionally given to lauding 'women's ways of knowing' as if these were different from, even superior to, logical or scientific ways of knowing. As Steven Pinker rightly said, such talk is an insult to women.

† Quoted in Carl Sagan, *The Demon-Haunted World*, p. 234. See also Paul R. Gross and Norman Levitt, *Higher Superstition*, for a chilling collection and justifiably savage indictment of similar drivel including 'Cultural Constructivism', 'Afrocentric Science', 'Feminist Algebra' and 'Science Studies', not forgetting Sandra Harding's 'stirring assertion that Newton's *Principia Mathematica Philosophae Naturalis* is a "rape manual"'.

‡ Winston Churchill, of course.

which culture you were brought up in, Bernoulli's Principle doesn't suddenly cease to operate as soon as you enter non-'Western' airspace. Or where do you put your money when it comes to predicting an observation? Like a modern Rider Haggard hero you can, as Carl Sagan pointed out, confound the savages of relativism and the New Age by predicting, to the second, a total eclipse of the sun a thousand years ahead.

Carl Sagan died a month ago. I met him once only but I have loved his books and I shall miss him as a 'candle in the dark'.* I dedicate this lecture to his memory, and shall use quotations from his writings. The remark about predicting eclipses is from the last book he published before he died, *The Demon-Haunted World*, and he goes on:

> You can go to the witch doctor to lift the spell that causes your pernicious anaemia, or you can take vitamin B12. If you want to save your child from polio, you can pray or you can inoculate. If you're interested in the sex of your unborn child, you can consult plumb-bob danglers all you want ... but they'll be right, on average, only one time in two. If you want real accuracy ... try amniocentesis and sonograms. Try science.

Of course, scientists often disagree with each other. But they are proud to agree on what new evidence it would take to change their minds. The route to any discovery will be published and whoever follows the same route should arrive at the same conclusion. If you lie – fiddle your figures, publish only that part of the evidence that supports your preferred conclusion – you will probably be found out. In any case, you won't get rich doing science, so why do it at all if you undermine the only point of the enterprise by lying? A scientist is much more likely to lie to a spouse or a tax inspector than to a scientific journal.

---

* I co-opted his phrase, together with Shakespeare's famous words from *Macbeth*, in the title of my second memoir, *Brief Candle in the Dark*.

Admittedly, there are cases of fraud in science, and probably more than come to light. My claim is only that in the scientific community fiddling data is the cardinal sin, unforgivable in a way that is hard to translate into the terms of another profession. An unfortunate consequence of this extreme value judgement is that scientists are exceptionally reluctant to blow the whistle on colleagues whom they may have reason to suspect of fiddling figures. It's rather like accusing somebody of cannibalism or paedophilia. Suspicions so dark may be suppressed until the evidence becomes too overwhelming to ignore, and by then much damage may have been done. If you fiddle your expense account, your peers will probably indulge you. If you pay a gardener in cash, thereby abetting a tax-dodging black market, you won't be a social pariah. But a scientist who is caught fiddling research data would. He would be shunned by his colleagues, and without mercy drummed out of the profession for ever.

A barrister who uses eloquence to make the best case he can, even if he doesn't believe it, even if he selects favourable facts and slants the evidence, would be admired and rewarded for his success.[*] A scientist who does the same thing, pulling out all the rhetorical stops, twisting and turning every way to win support for a favourite theory, is regarded with at least mild suspicion.

[*] The following experience is commonplace. I was once talking to a barrister, a young woman of high ideals specializing in criminal law defence. She expressed satisfaction that a private investigator whom she had employed had found evidence exonerating her client, who was accused of murder. I congratulated her, and asked the obvious question: what would she have done if he had found evidence proving unequivocally that her client was guilty? Without hesitation she said that she would have quietly suppressed the evidence. Let the prosecution find their own evidence. If they failed, more fool them. My outraged reaction to this story was one that she had obviously met many times when talking to non-lawyers, and I didn't blame her for wearily changing the subject rather than pursuing the argument.

Typically, the values of scientists are such that the charge of advocacy – or, worse, of being a *skilled* advocate – is a charge that needs to be answered.* But there is an important difference between using rhetoric to bring out what you believe is really there, and using rhetoric knowingly to cover up what is really there. I once spoke in a university debate on evolution. The most effective creationist speech was made by a young woman who happened to be placed next to me at dinner afterwards. When I complimented her on her speech, she immediately told me she hadn't believed a word of it. She was simply exercising her debating skills by arguing passionately for the exact opposite of what she considered to be true. No doubt she will make a good lawyer. The fact that now it was all I could do to stay polite to my dinner companion may say something about the values that I have acquired as a scientist.

I suppose I am saying that scientists have a scale of values according to which there is something almost sacred about nature's truth. This may be why some of us get so heated about astrologers, spoonbenders and similar charlatans, whom others indulgently tolerate as harmless entertainers. The law of libel penalizes those who knowingly tell lies about individuals. But you get off scot-free if you make money lying about nature – who can't sue. My values may be warped, but I'd like nature to be represented in court like an abused child.†

---

* I felt it necessary to begin *The Extended Phenotype* by admitting that it was a work of 'unabashed advocacy'. The fact that I needed to use a word like 'unabashed' speaks to my point about the values of science. What lawyer would apologize to the jury for his 'unabashed advocacy'? Advocacy, partisan advocacy, is precisely what lawyers are trained – and handsomely paid – for. Politicians, too, and advertising or marketing professionals. Science is perhaps the most stringently honest of all professions.

† I heard of a London physicist who went to the lengths of refusing to pay his local government tax as long as the area's adult education college advertised a course in astrology. An Australian professor

The downside to the love of truth is that it may lead scientists to pursue it regardless of unfortunate consequences.* Scientists do bear a heavy responsibility to warn society of those consequences. Einstein acknowledged the danger when he said: 'If I had only known, I would have been a locksmith.' But of course he wouldn't really. And when the opportunity came, he signed the famous letter alerting Roosevelt to the possibilities and dangers of the atomic bomb. Some of the hostility meted out to scientists is equivalent to shooting the messenger. If astronomers called our attention to a large asteroid on a collision course for Earth, the final thought of many people before impact would be to blame 'the scientists'. There is an element of shooting the messenger about our reaction to BSE.†
Unlike the asteroid case, here the true blame does belong with

---

of geology is in the process of suing a creationist for making money under false pretences by claiming to have found Noah's Ark. See commentary by Peter Pockley, *Daily Telegraph*, 23 April 1997.

* I'd find it hard to justify funding research on alleged correlations between race and IQ. I'm not one of those who thinks that intelligence is unmeasurable or that race is 'non-biological', a 'social construct' (see the distinguished geneticist A. W. F. Edwards' splendid take-down of this claim in 'Human genetic diversity: Lewontin's fallacy'). But what could possibly be the point of investigating alleged correlations between intelligence and race? Certainly no policy decisions should ever be based on such research. That, I suspect, was the point Lewontin really intended to make and I unreservedly agree. However, as so often with ideologically motivated scientists, he chose to misrepresent his point as a (false) scientific one rather than as a (laudable) political one.

† Bovine spongiform encephalopathy, commonly known as 'mad cow disease'. An epidemic in Britain starting in 1986 caused widespread alarm, partly because of its affinity to the dangerous human malady CJD or Creutzfeldt–Jakob Disease.

humanity. Scientists must bear some of it, along with the economic greed of the agricultural foodstuffs industry.

Carl Sagan remarks that he is often asked whether he believes there is intelligent life out there. He leans towards a cautious yes, but says it with humility and uncertainty.

> Often, I'm asked next, 'What do you really think?'
> I say, 'I just told you what I really think.'
> 'Yes, but what's your gut feeling?'
> But I try not to think with my gut. If I'm serious about understanding the world, thinking with anything besides my brain, as tempting as that might be, is likely to get me into trouble. Really, it's okay to reserve judgement until the evidence is in.

Mistrust of inner, private revelation is, it seems to me, another of the values fostered by the experience of doing science. Private revelation doesn't sit well with the textbook ideals of scientific method: testability, evidential support, precision, quantifiability, consistency, intersubjectivity, repeatability, universality, and independence of cultural milieu.

There are also values of science which are probably best treated as akin to aesthetic values. Einstein on the subject is sufficiently often quoted so here, instead, is the great Indian astrophysicist Subrahmanyan Chandrasekhar, in a lecture in 1975, when he was sixty-five:

> In my entire scientific life . . . the most shattering experience has been the realisation that an exact solution of Einstein's equations of general relativity, discovered by the New Zealand mathematician Roy Kerr, provides the absolutely exact representation of untold numbers of massive black holes that populate the Universe. This 'shuddering before the beautiful', this incredible fact that a discovery motivated by a search after the beautiful in mathematics should find its exact replica in Nature, persuades me to say that beauty is that to which the human mind responds at its deepest and most profound.

I find this moving in a way that is missing from the skittish dilettantism of Keats' famous lines:

> 'Beauty is truth, truth beauty,' – that is all
> Ye know on earth, and all ye need to know.

Going only a little beyond aesthetics, scientists tend to value the long term at the expense of the short; they draw inspiration from the wide open spaces of the cosmos and the grinding slowness of geological time rather than the parochial concerns of humanity. They are especially prone to see things *sub specie aeternitatis*, even if this puts them at risk of being accused of a bleak, cold, unsympathetic view of humanity.

Carl Sagan's penultimate book, *Pale Blue Dot*, is built around the poetic image of our world seen from distant space.

> Look again at that dot. That's here. That's home . . . The Earth is a very small stage in a vast cosmic arena. Think of the rivers of blood spilled by all those generals and emperors so that, in glory and triumph, they could become the momentary masters of a fraction of a dot. Think of the endless cruelties visited by the inhabitants of one corner of this pixel on the scarcely distinguishable inhabitants of some other corner, how frequent their misunderstandings, how eager they are to kill one another, how fervent their hatreds.
>
> Our posturings, our imagined self-importance, the delusion that we have some privileged position in the Universe, are challenged by this point of pale light. Our planet is a lonely speck in a great enveloping cosmic dark. In our obscurity, in all this vastness, there is no hint that help will come from elsewhere to save us from ourselves.

For me the only bleak aspect of the passage I have just read is the human realization that its author is now silenced. Whether the scientific cutting down to size of humanity seems bleak is a matter of attitude. It may be an aspect of scientific values that many of us find such large visions uplifting and exhilarating rather than cold and empty. We also warm to nature as lawful and uncapricious.

There is mystery but never magic, and mysteries are all the more beautiful for being eventually explained. Things are explicable and it is our privilege to explain them. The principles that operate here prevail there – and 'there' means out to distant galaxies. Charles Darwin, in the famous 'entangled bank' passage which ends *The Origin of Species*, notes that all the complexity of life has 'been produced by laws acting around us . . .' and he goes on:

> Thus, from the war of nature, from famine and death, the most exalted object which we are capable of conceiving, namely, the production of the higher animals, directly follows. There is grandeur in this view of life, with its several powers, having been originally breathed into a few forms or into one; and that, whilst this planet has gone cycling on according to the fixed law of gravity, from so simple a beginning endless forms most beautiful and most wonderful have been, and are being, evolved.

The sheer time it has taken species to evolve constitutes a favoured argument for their conservation. This in itself involves a value judgement, presumably one congenial to those steeped in the depths of geological time. In a previous work I have quoted Oria Douglas-Hamilton's harrowing account of an elephant cull in Zimbabwe:

> I looked at one of the discarded trunks and wondered how many millions of years it must have taken to create such a miracle of evolution. Equipped with fifty thousand muscles and controlled by a brain to match such complexity, it can wrench and push with tonnes of force . . . at the same time, it is capable of performing the most delicate operations . . . And yet there it lay, amputated like so many elephant trunks I had seen all over Africa.

Moving as this is, I quote it to illustrate the scientific values that led Mrs Douglas-Hamilton to stress the millions of years it has taken to evolve the complexity of an elephant's trunk rather than, say, the rights of elephants or their capacity to suffer, or the value of wildlife in enriching our human experience or a country's tourist revenues.

Not that evolutionary understanding is irrelevant to questions of rights and suffering. I am shortly going to support the view that you cannot derive fundamental moral values from scientific knowledge. But utilitarian moral philosophers who do not believe there *are* any absolute moral values nevertheless justly claim a role in unmasking contradictions and inconsistencies within particular value systems.* Evolutionary scientists are well placed to observe inconsistencies in the absolutist elevation of human rights over those of all other species.

'Pro-lifers' assert, without question, that life is infinitely precious, while cheerfully tucking into a large steak. The sort of 'life' that such people are 'pro' is all too clearly *human* life. Now, this is not necessarily wrong, but the evolutionary scientist will, at very least, warn us of inconsistency. It is not self-evident that abortion of a one-month human foetus is murder, while shooting a fully sentient adult elephant or mountain gorilla is not.

Some six or seven million years ago there lived an African ape which was the common ancestor of all modern humans and all modern gorillas. By chance, the intermediate forms that link us to this ancestor – *Homo erectus, Homo habilis*; various members of the genus *Australopithecus* and others – are extinct. Also extinct are the intermediates that link the same common ancestor to modern gorillas. If the intermediates were not extinct; if relict populations turned up in the jungles and savannahs of Africa; the consequences would be poignant. You'd be able to mate and have a child with someone who'd be able to mate and have a child with someone else who . . . after a handful of further links in the chain, would be able

* My favourite moral philosopher, and an excellent example of how valuable philosophers can be when they aim for clarity without pretentious languaging-up, is Jonathan Glover. See, for example, his *Causing Death and Saving Lives*, a book so far-sighted that it was allowed to go out of print before scientific advances started to make it really topical; or his *Humanity*, which turns out to be a searing indictment of the opposite. In *Choosing Children*, which ventures into the near-taboo subject of eugenics, Glover demonstrates the intellectual courage that goes with the territory of honest moral philosophy.

to mate and have a child with a gorilla. It is sheer bad luck that some key intermediates in this chain of interfertility happen to be dead.

This is not a frivolous thought experiment. The only room for argument is over how many intermediate stages we need to postulate in the chain. And it doesn't matter how many intermediate stages there are in order to justify the following conclusion. Your absolutist elevation of *Homo sapiens* above all other species, your unargued preference for a human foetus or a brain-dead human vegetable over an adult chimpanzee at the height of its powers, your species-level apartheid, would collapse like a house of cards. Or if it did not, the comparison with apartheid would turn out to be no idle one. For if, in the face of a surviving continuum of intermediates, you insisted upon separating humans from non-humans, you could maintain the separation only by appealing to apartheid-like courts to decide whether particular intermediate individuals could 'pass for human'.

Such evolutionary logic does not destroy all doctrines of specifically human rights. But it certainly destroys absolutist versions, for it shows that the separation of our species depends upon accidents of extinction. If morals and rights were absolute in principle, they could not be jeopardized by new zoological discoveries in the Budongo Forest.

Values of science in the strong sense

I want to turn now from the weak to the strong sense of the values of science, to scientific findings as the direct source of a system of values. The versatile English biologist Sir Julian Huxley – incidentally a predecessor of mine as Tutor in Zoology at New College – tried to make evolution the basis for an ethics, almost for a religion. For him, The Good is that which furthers the evolutionary process. His more distinguished but unknighted grandfather Thomas Henry Huxley took an almost opposite view. I am more in sympathy with Huxley senior.*

---

* Julian Huxley edited a compilation of his own and his grandfather's views on the subject, entitled *Touchstone for Ethics*.

Part of Julian Huxley's ideological infatuation with evolution stemmed from his optimistic vision of its *progress*.* Nowadays it is fashionable to doubt that evolution really is progressive at all. This is an interesting argument and I have a view on it,[†] but it is superseded by the prior question of whether we should anyway base our values upon this or any other conclusion about nature.

A similar point arises about Marxism. You can espouse an academic theory of history which predicts the dictatorship of the proletariat. And you can follow a political creed which values the dictatorship of the proletariat as a good thing that you should work to encourage. Many Marxists as a matter of fact do both, and a disconcerting number, arguably including Marx himself, cannot

---

* His 'Progress, biological and other', the first of his *Essays of a Biologist*, contains passages that read almost like a call to arms under evolution's banner: '[man's] face is set in the same direction as the main tide of evolving life, and his highest destiny, the end towards which he has so long perceived that he must strive, is to extend to new possibilities the process with which, for all these millions of years, nature has already been busy, to introduce less and less wasteful methods, to accelerate by means of his consciousness what in the past has been the work of blind unconscious forces' (page 41). That passage exemplifies what I shall disparage, on page 150, as 'poetic science' – poetic in the bad sense, not the good sense implied by my title *Science in the Soul*. Huxley's book of essays influenced me deeply when I read it as an undergraduate. I am now less impressed, and rather subscribe to the view which I once heard Peter Medawar utter in an audaciously unguarded moment: 'The trouble with Julian is that he just doesn't understand evolution!'

† Stephen J. Gould, in *Full House*, rightly attacked 'progress' where it is taken to mean progress towards the lofty evolutionary peak of humanity. But in my critical review of his book, published in *Evolution* (1997), I defend 'progress' where it means evolutionary movement in consistently the same direction towards the build-up of adaptive complexity, the drive often being powered by 'evolutionary arms races'.

tell the difference. But logically, the political belief in what is desirable does not follow from the academic theory of history. You could consistently be an academic Marxist believing that the forces of history drive inexorably towards a workers' revolution, while at the same time voting High Tory and working as hard as possible to postpone the inevitable. Or you could be a passionate Marxist politically, who works all the harder for the revolution precisely because you doubt the Marxist theory of history and feel that the longed-for revolution needs all the help it can get.

Similarly, evolution may or may not have the quality of progressiveness that Julian Huxley, as an academic biologist, supposed. But whether he was right or not about the biology, it clearly is not necessary that we should imitate this kind of progressiveness in drawing up our systems of values.

The issue is even starker if we move from evolution itself, with its alleged progressive momentum, to Darwin's mechanism of evolution, the survival of the fittest. T. H. Huxley, in his 1893 Romanes Lecture, 'Evolution and Ethics', was under no illusions, and he was right. If you must use Darwinism as a morality play, it is an awful warning. Nature really is red in tooth and claw. The weakest really do go to the wall, and natural selection really does favour selfish genes. The racing elegance of cheetahs and gazelles is bought at huge cost in blood and the suffering of countless antecedents on both sides. Ancient antelopes were butchered and carnivores starved, in the shaping of their streamlined modern counterparts. The product of natural selection, life in all its forms, is beautiful and rich. But the process is vicious, brutal and short-sighted.

As an academic fact we are Darwinian creatures, our forms and our brains sculpted by natural selection, that indifferent, cruelly blind watchmaker. But this doesn't mean we have to like it. On the contrary, a Darwinian society is not the sort of society in which any friend of mine would wish to live. 'Darwinian' is not a bad definition of precisely the sort of politics I would run a hundred miles not to be governed by, a sort of over-the-top Thatcherism gone native.

I should be allowed a personal word here because I am tired of being identified with a vicious politics of ruthless competitiveness, accused of advancing selfishness as a way of life. Soon after Mrs Thatcher's election victory of 1979, Professor Steven Rose wrote, in *New Scientist*, as follows:

> I am not implying that Saatchi and Saatchi engaged a team of sociobiologists to write the Thatcher scripts, nor even that certain Oxford and Sussex dons are beginning to rejoice at this practical expression of the simple truths of selfish genery they have been struggling to convey to us. The coincidence of fashionable theory with political events is messier than that. I do believe though, that when the history of the move to the right of the late 1970s comes to be written, from law and order to monetarism and to the (more contradictory) attack on statism, then the switch in scientific fashion, if only from group to kin selection models in evolutionary theory, will come to be seen as part of the tide which has rolled the Thatcherites and their concept of a fixed, 19th century competitive and xenophobic human nature into power.

The 'Sussex don' referred to was John Maynard Smith, and he gave the apt reply in a letter to the next issue of *New Scientist*: What should we have done, fiddled the equations?

Rose was a leader of the Marxist-inspired attack of the time on sociobiology. It is entirely typical that, just as these Marxists were incapable of separating their academic theory of history from their normative political beliefs, they assumed that we were incapable of separating our biology from our politics. They simply could not grasp that one might hold academic beliefs about the way evolution happens in nature, while simultaneously repudiating the desirability of translating those academic beliefs into politics. This led them to the untenable conclusion that, since genetic Darwinism when applied to humans had undesirable political connotations, it must not be *allowed* to be scientifically correct.*

---

* I made the same point with reference to Rose's Marxist co-author Richard Lewontin in an earlier footnote to this essay (see page 27 above).

They and many others make the same kind of mistake with respect to positive eugenics. The premiss is that to breed humans selectively for abilities such as running speed, musical talent or mathematical dexterity would be politically and morally indefensible. Therefore it isn't (*must* not be) possible – ruled out by science. Well, anybody can see that that's a *non sequitur*, and I'm sorry to have to tell you that positive eugenics is not ruled out by science. There is no reason to doubt that humans would respond to selective breeding just as readily as cows, dogs, cereal plants and chickens. I hope it isn't necessary for me to say that this doesn't mean I am in favour of it.

There are those who will accept the feasibility of physical eugenics but dig their trench before mental eugenics. Maybe you could breed a race of Olympic swimming champions, they concede, but you will never breed for higher intelligence, either because there's no agreed method of measuring intelligence, or because intelligence is not a single quantity varying in one dimension, or because intelligence doesn't vary genetically, or some combination of these three points.

If you seek refuge in any of these lines of thought, it is once again my unpleasant duty to disillusion you. It doesn't matter if we can't agree how to measure intelligence, we can breed for any of the disputed measures, or a combination of them. It might be hard agreeing a definition for docility in dogs, but this doesn't stop us breeding for it. It doesn't matter if intelligence is not a single variable, the same is probably true of milking prowess in cows and racing ability in horses. You can still breed for them, even while disputing how they should be measured, or whether they each constitute a single dimension of variation.

As for the suggestion that intelligence, measured in any way or in any combination of ways, does not vary genetically, it more or less cannot be true, for the following reason whose logic requires only the premiss that we are more intelligent – by whatever definition you choose – than chimpanzees and all other apes. If we are more intelligent than the ape which lived six million years ago and was our common ancestor with chimpanzees, there has been an evolutionary trend in our ancestry towards increased intelligence. There has certainly been an evolutionary trend towards increased brain size: it is one of the more dramatic evolutionary trends in the vertebrate fossil record. Evolutionary trends cannot happen unless there is genetic variation in the characteristics concerned – in this case brain size and presumably intelligence. So, there was genetic variation in intelligence in our ancestors. It is just possible that there isn't any longer, but such an exceptional circumstance would be bizarre. Even if the evidence from twin studies* did not support it – which it does – we could safely draw the conclusion, from evolutionary logic alone, that we have genetic variance in intelligence, intelligence being defined in terms of whatever separates us from our ape ancestors. Using the same definition, we could, if we wanted to, use artificial selective breeding to continue the same evolutionary trend.

I would need little persuading that such a eugenic policy would be politically and morally wrong,† but we must be absolutely clear

---

* Twin studies constitute a powerful and easily understood technique for estimating the contribution of genes to variance. Measure something (anything you like) in pairs of monozygotic twins (who are known to be genetically identical). Compare their similarity (to each other) with the similarity (to each other) of dizygotic twins (who are no more likely to share genes than ordinary siblings). If the pairs of monozygotic twins resemble each other significantly more than the pairs of dizygotic twins in, for example, intelligence, you can conclude that genes are responsible. The twin-study technique is especially persuasive in those rare – and much studied – cases where monozygotic twins happen to be separated at birth and brought up apart.

† Any kind of government-imposed eugenic policy, breeding positively for some nationally sought-after characteristic such as running speed or intelligence, would be a lot harder to justify than a voluntary version. In vitro fertilization (IVF) techniques hormonally stimulate women to super-ovulate, producing as many as a dozen eggs. Of those that are successfully fertilized in

that such a *value judgement* is the right reason to refrain from it. Let us not allow our value judgements to push us over into the false scientific belief that human eugenics isn't *possible*. Nature, fortunately or unfortunately, is indifferent to anything so parochial as human values.

Later, Rose joined forces with Leon Kamin, one of America's leading opponents of IQ-measuring, and with the distinguished Marxist geneticist Richard Lewontin, to write a book in which they repeated these and many other errors.* They also acknowledged that the sociobiologists wanted to be less fascist than our science, in their (mistaken) view, ought to make us, but they (equally mistakenly) tried to catch us in a contradiction with the mechanistic interpretation of mind which we – and presumably they – follow.

---

the petri dish, only two or perhaps three are reinserted in the woman, in the hope that one of them will 'take'. The choice is normally made at random. However, it is possible for one cell from an eight-cell conceptus to be extracted without damage, and the genes assayed. This means that the choice of which ones are reinserted, which ones discarded, can be non-random with respect to genes. Few people would object to this technique being used to select against a condition such as haemophilia or Huntington's disease – 'negative eugenics'. However, many recoil from using the same technique in 'positive eugenics': selecting, in the petri dish, for musical ability, say, if that were one day to become possible. Yet the same people don't object to ambitious parents imposing music lessons and piano practice on their children. There may be good reasons for this double standard, but they need to be discussed. It is at the very least important to distinguish voluntary eugenics practised by individual parents from state-imposed eugenics such as the Nazis brutally implemented.

* S. Rose, L. J. Kamin and R. C. Lewontin, *Not in our Genes*. The order of authors, weirdly, is different from that in the American edition, in which Rose and Lewontin change places. My review of the book in *New Scientist*, vol. 105, 1985, pp. 59–60, gives a full critique which briefly earned me, and *New Scientist*, the threat of a lawsuit. I stand by every word of it.

> Such a position is, or ought to be, completely in accord with the principles of sociobiology offered by Wilson[*] and Dawkins. However, to adopt it would involve them in the dilemma of first arguing the innateness of much human behavior that, being liberal men, they clearly find unattractive (spite, indoctrination, etc.) ... To avoid this problem, Wilson and Dawkins invoke a free will that enables us to go against the dictates of our genes if we so wish.

This, they complain, is a return to unabashed Cartesian dualism. You cannot, say Rose and his colleagues, believe that we are survival machines programmed by our genes, and at the same time urge rebellion against them.

What's the problem? Without going into the difficult philosophy of determinism and free will,[†] it is easy to observe that, as a matter of fact, we do go against the dictates of our genes. We rebel every time we use contraception when we'd be economically capable of rearing a child. We rebel when we give lectures, write books or compose sonatas instead of single-mindedly devoting our time and energy to disseminating our genes.

This is easy stuff; there is no philosophical difficulty at all. Natural selection of selfish genes gave us big brains which were originally useful for survival in a purely utilitarian sense. Once those big brains, with their linguistic and other capacities, were in

---

[*] Edward O. Wilson, author of *Sociobiology*.

[†] For a view of this topic which many scientists will find congenial, see Daniel C. Dennett's *Elbow Room*. Dennett has returned to the question in later books such as *Freedom Evolves* and *From Bacteria to Bach and Back*. Not all scientists and philosophers agree with Dennett's version of 'compatibilism', however. Jerry Coyne and Sam Harris are among those who do not. After my public speeches I have come to dread the near-inevitable 'Do you believe in free will?' question, and sometimes resort to quoting Christopher Hitchens' characteristically witty answer: 'I have no choice.' What I will more confidently say, in reply to Rose and Lewontin, is that the addition of the word 'genetic' before 'determinism' doesn't make it any more deterministic.

place, there is no contradiction at all in saying that they took off in wholly new 'emergent' directions, including directions opposed to the interests of selfish genes.

There is nothing self-contradictory about emergent properties. Electronic computers, conceived as calculating machines, emerge as word processors, chess players, encyclopedias, telephone switchboards, even, I regret to say, electronic horoscopes. No fundamental contradictions are there to ring philosophical alarm bells. Nor in the statement that our brains have overtaken, even overreached, their Darwinian provenance. Just as we defy our selfish genes when we wantonly detach the enjoyment of sex from its Darwinian function, so we can sit down together and with language devise politics, ethics and values which are vigorously anti-Darwinian in their thrust. I shall return to this in my conclusion.

One of Hitler's perverted sciences was a garbled Darwinism and, of course, eugenics. But, uncomfortable though it is to admit it, Hitler's views were not unusual in the first part of this century. I quote from a chapter on 'The New Republic', an allegedly Darwinian utopia, written in 1902:

> And how will the New Republic treat the inferior races? How will it deal with the black? how will it deal with the yellow man? . . . those swarms of black, and brown, and dirty-white, and yellow people, who do not come into the new needs of efficiency? Well, the world is a world, and not a charitable institution, and I take it they will have to go . . . And the ethical system of these men of the New Republic, the ethical system which will dominate the world state, will be shaped primarily to favour the procreation of what is fine and efficient and beautiful in humanity – beautiful and strong bodies, clear and powerful minds.

The author of this is not Adolf Hitler but H. G. Wells,* who thought of himself as a socialist. It is stuff like this (and there's lots more

* In *Anticipations of the Reaction of Mechanical and Scientific Progress upon Human Life and Thought*. My lecture included a longer quotation from Wells's book.

from the Social Darwinists) that has given Darwinism a bad name in the social sciences. And how! But, again, we must not attempt to use the facts of nature to derive our politics or our morality one way or the other. David Hume is to be preferred to either of the two Huxleys: moral directives cannot be derived from descriptive premises, or, as it is more colloquially put, 'You can't get an "ought" from an "is"'. Where then, on the evolutionary view, do our 'oughts' come from? Where do we get our values, moral and aesthetic, ethical and political? It is time to move on from the values of science to the science of values.

The science of values

Have we inherited our values from remote ancestors? The onus is on those who would deny it. The tree of life, Darwin's tree, is a vast, bushy thicket of thirty million twigs.* We are one tiny twig, buried somewhere in the surface layers. Our twig sprouts from a small bough alongside our ape cousins, not far from the larger bough of our monkey cousins, within view of our more distant cousins, cousin kangaroo, cousin octopus, cousin staphylococcus. Nobody doubts that all the rest of the thirty million twigs inherit their

---

* This is the highest estimate I have seen for living species. The true figure is unknown and may be substantially lower, but if you include extinct species it is certainly higher. To draw a tree diagram of the complete pedigree of all life you'd need a sheet of paper whose acreage would cover the island of Manhattan six times over. James Rosindell was accordingly moved to write the brilliant 'OneZoom' software, which represents the entire tree of life as a fractal. You can fly over it on your computer screen like a sort of taxonomic Google Earth, and 'drill down' to any particular species you fancy. OneZoom is now being fleshed out in collaboration with Yan Wong, my co-author of *The Ancestor's Tale*, the second edition of which makes extensive use of it. Rosindell and Wong invite enthusiasts (I am one) to sponsor favourite species to defray the costs of adding their details to the tree.

attributes from their ancestors, and by any standards we humans owe to our ancestors much of what we are and what we look like. We have inherited from our forebears – with greater or less modification – our bones and eyes, our ears and thighs, even, it is hard to doubt, our lusts and our fears. *A priori* there seems no obvious reason why the same should not apply to our higher mental faculties, our arts and our morals, our sense of natural justice, our values. Can we exclude these manifestations of high humanity from what Darwin called the indelible stamp of our lowly origins? Or was Darwin right when he remarked more informally to himself in one of his notebooks, 'He who understands baboon would do more towards metaphysics than Locke'? I shall make no attempt to review the literature, but the question of the Darwinian evolution of values and morals has been frequently and extensively discussed.

Here's the fundamental logic of Darwinism. Everybody has ancestors but not everybody has descendants. We have all inherited the genes for being an ancestor, at the expense of the genes for failing to be an ancestor. Ancestry is the ultimate Darwinian value. In a purely Darwinian world, all other values are subsidiary. Synonymously, gene survival is the ultimate Darwinian value. As a first expectation, all animals and plants can be expected to work ceaselessly for the long-term survival of the genes that ride inside them.

The world is divided into those for whom the simple logic of this is as clear as daylight, and those who, no matter how many times it is explained to them, just don't get it. Alfred Wallace wrote about the problem* in a letter to his co-discoverer of natural selection: 'My dear Darwin – I have been so repeatedly struck by the utter inability of numbers of intelligent persons to see clearly, or at all, the self-acting and necessary effects of natural selection . . .'

Those who don't get it either assume that there must be some kind of personal agent in the background to do the choosing, or they wonder why individuals should value survival of their own

---

\| * In nineteenth-century terms without reference to genes, of course.

genes, rather than, for instance, the survival of their species, or the survival of the ecosystem of which they are a part. After all, say this second group of people, if the species and the ecosystem don't survive, nor will the individual, so it is in their interests to value the species and the ecosystem. Who decides, they wonder, that gene survival is the ultimate value?

Nobody decides. It follows automatically from the fact that genes reside in the bodies that they build and are the only things (in the form of coded copies) that can persist from one generation of bodies to the next. This is the modern version of the point Wallace was making with his apt phrase 'self-acting'. Individuals are not miraculously or cognitively inspired with values and goals that will guide them in the paths of gene survival. Only the past can have an influence, not the future. Animals behave *as if* striving for the future values of the selfish gene simply and solely because they bear, and are influenced by, genes that survived through ancestral generations in the past. Those ancestors that, in their own time, behaved as if they valued whatever was conducive to the future survival of their genes, have bequeathed those very genes to their descendants. So their descendants behave as if they, in their turn, value the future survival of their genes.

It is an entirely unpremeditated, self-acting process which works so long as future conditions are tolerably similar to past. If they are not, it doesn't, and the result is often extinction. Those who understand this understand Darwinism. The word Darwinism, by the way, was coined by the ever-generous Wallace. I shall continue my Darwinian analysis of values using bones as my example, because they are unlikely to ruffle human hackles and therefore distract.

Bones are not perfect; they sometimes break. A wild animal that breaks its leg is unlikely to survive in the harsh, competitive world of nature. It will be especially vulnerable to predators, or unable to catch prey. So why doesn't natural selection thicken bones so that they never break? We humans, by artificial selection, could breed a race of, say, dogs, whose leg bones were so stout that they never broke. Why doesn't nature do the same? Because of costs, and this implies a system of values.

Engineers and architects are never asked to build unbreakable structures, impregnable walls. Instead, they are given a monetary budget and asked to do the best they can, according to certain criteria, within that constraint. Or they may be told: the bridge must bear a weight of ten tons, and must withstand gales three times more forceful than the worst ever recorded in this gorge. Now design the most economical bridge you can that meets these specifications. Safety factors in engineering imply monetary valuation of human life. Designers of civilian airliners are more risk-averse than designers of military aircraft. All aircraft and ground control facilities could be safer if more money was spent. More redundancy could be built into control systems, the number of flying hours demanded of a pilot before he is allowed to carry live passengers could be increased. Baggage inspection could be more stringent and time-consuming.

The reason we don't take these steps to make life safer is largely one of cost. We are prepared to pay a lot of money, time and trouble for human safety, but not infinite amounts. Like it or not, we are forced to put monetary value on human life. In most people's scale of values, human life rates higher than non-human animal life, but animal life does not have zero value. Notoriously, the evidence of newspaper coverage suggests that people value life belonging to their own race higher than human life generally. In wartime, both absolute and relative valuations of human life change dramatically. People who think it is somehow wicked to talk about this monetary valuation of human life – people who emotionally declare that a single human life has infinite value – are living in cloud-cuckoo-land.

Darwinian selection, too, optimizes within economic limits and can be said to have values in the same sense. John Maynard Smith said: 'If there were no constraints on what is possible, the best phenotype would live for ever, would be impregnable to predators, would lay eggs at an infinite rate, and so on.'

Nicholas Humphrey continues the argument with another analogy from engineering.

> Henry Ford, it is said,[*] commissioned a survey of the car scrap yards of America to find out if there were parts of the Model T Ford which never failed. His inspectors came back with reports of almost every kind of breakdown: axles, brakes, pistons – all were liable to go wrong. But they drew attention to one notable exception: the *kingpins* of the scrapped cars invariably had years of life left in them. With ruthless logic Ford concluded that the kingpins on the Model T were too good for their job and ordered that in future they should be made to an inferior specification . . . Nature is surely at least as careful an economist as Henry Ford.

Humphrey applied his lesson to the evolution of intelligence, but it can equally be applied to bones or anything else. Let us commission a survey of the corpses of gibbons, to see if there are some bones that never fail. We find that every bone in the body breaks at one time or another, but with one notable exception: let's (rather implausibly) say the femur (thigh bone) is never known to break. Henry Ford would be in no doubt. In future, the femur must be made to an inferior specification.

Natural selection would agree. Individuals with slightly thinner femurs, who diverted the material saved into some other purpose, say building up other bones and making them less likely to break, would survive better. Or females might take the calcium shaved off the thickness of the femur and put it into milk, thereby improving

---

* Said by whom? Nobody seems to know. The suspicion that the answer might be Nicholas Humphrey himself doesn't threaten the pertinence of his parable. And Ford himself would probably not have minded. I've quoted Humphrey's tale so often that my friend the enigmatically humorous ichthyologist David Noakes went to the trouble of procuring, and sending to me out of the blue, the kingpin of a Model T, which looks, I must say, in pristine condition and weighty enough to seem overdesigned.

the survival of their offspring – and with them the genes for making the economy.

In a machine or an animal the (simplified) ideal is that all the parts should wear out simultaneously. If there is one part that consistently has years of life left in it after the others have worn out, it is overdesigned. Materials that went into building it up should, instead, be diverted to other parts. If there is one part that consistently wears out before anything else, it is underdesigned. It should be built up, using materials taken away from other parts. Natural selection will tend to uphold an equilibration rule: 'Rob from strong bones to pay weak ones, until all are of equal strength.'

The reason this is an oversimplification is that not all the bits of an animal or a machine are equally important. That's why inflight entertainment systems go wrong thankfully more often than rudders or jet engines. A gibbon might be able to afford a broken femur better than a broken humerus. Its way of life depends upon 'brachiation' (swinging through the trees by its arms). A gibbon with a broken leg might survive to have another child. A gibbon with a broken arm probably wouldn't. So the equilibration rule I mentioned has to be tempered: 'Rob from strong bones to pay weak ones, until you have equalized the risks to your survival accruing from breakages in all parts of your skeleton.'

But who is the 'you' admonished in the equilibration rule? It certainly isn't an individual gibbon, who is not, we assume, capable of making compensatory adjustments to its own bones. The 'you' is an abstraction. You can think of it as a lineage of gibbons in ancestor/descendant relation to one another, represented by the genes that they share. As the lineage progresses, ancestors whose genes make the right adjustments survive to leave descendants who inherit those correctly equilibrating genes. The genes that we see in the world tend to be the ones that get the equilibration right, because they have survived through a long line of successful ancestors who have not suffered the breakage of underdesigned, or the waste of overdesigned, bones.

46

So much for bones. Now we need to establish, in Darwinian terms, what *values* are doing for animals and plants. Where bones stiffen limbs, what do values do for their possessors? By values, I am now going to mean the criteria, in the brain, by which animals choose how to behave.

The majority of things in the universe don't actively strive for anything. They just are. I am concerned with the minority that do strive for things, entities that appear to work towards some end and then stop when they've achieved it. This minority I shall call value-driven. Some of them are animals and plants, some man-made machines.

Thermostats, heat-seeking Sidewinder missiles, and numerous physiological systems in animals and plants are controlled by negative feedback. There is a target value which is defined in the system. Discrepancies from the target value are sensed and fed back into the system, causing it to change its state in the direction of reducing the discrepancy.

Other value-seeking systems improve with experience. From the point of view of defining values in learning systems, the key concept is *reinforcement*. Reinforcers are positive ('rewards') or negative ('punishments'). Rewards are states of the world which, when encountered, cause an animal to repeat whatever it recently did. Punishments are states of the world which, when encountered, cause an animal to avoid repeating whatever it recently did.

The stimuli that animals treat as rewards and punishments can be seen as values. Psychologists make a further distinction between primary and secondary reinforcers (both rewards and punishments). Chimpanzees learn to work for food as a primary reward, but they will also learn to work for the equivalent of money – secondary rewards – plastic tokens which they have previously learned to stuff into a slot machine to get food.

Some psychological theorists have argued that there is only one primary built-in reward ('drive reduction' or 'need reduction'), upon which all others are built. Others, such as Konrad Lorenz, the grand old man of ethology,* argued that Darwinian natural selection has built-in complicated rewarding mechanisms, specified differently and in detail for each species to fit its unique way of life.

Perhaps the most elaborately detailed examples of primary values come from bird song. Different species develop their songs in different ways. The American song sparrow is a fascinating mixture. Young birds brought up completely alone end up singing normal song sparrow song. So, unlike, say, bullfinches, they don't learn by imitation. But they do learn. Young song sparrows teach themselves to sing by babbling at random and repeating those fragments that match a built-in template. The template is a genetically specified preconception of what a song sparrow ought to sound like. You could say that the information is built in by the genes, but into the sensory part of the brain. It has to be transferred to the motor side by learning. And the sensation specified by the template, by definition, is a reward: the bird repeats actions that deliver it. But, as rewards go, it is very elaborate and precisely specified in detail.

It is examples like this that stimulated Lorenz to use the colourful phrase 'innate schoolmarm' (or 'innate teaching mechanism') in his lengthy attempts to resolve the ancient dispute over nativism versus environmentalism. His point was that, however important learning is, there has to be innate guidance of what we shall learn. In particular, each species needs to be supplied with its own specifications of what to treat as rewarding, what punishing. *Primary* values, Lorenz was saying, have to come from Darwinian natural selection.

Given enough time we should be able to breed, by artificial selection, a race of animals that enjoy pain and hate pleasure. Of course, by the animals' newly evolved definition this statement is

---

* With his handsomely patrician head of hair and matching white beard, it is said (there we go again; see footnote on page 45) that he exploited his resemblance to God when soliciting charitable donations from rich old ladies.

oxymoronic. I'll rephrase it. By artificial selection we could reverse the previous definitions of pleasure and pain.*

The animals so modified will be less well equipped to survive than their wild ancestors. Wild ancestors have been naturally selected to enjoy those stimuli most likely to improve their survival, and to treat as painful those stimuli most likely, statistically, to kill them. Injury to the body, puncturing skin, breaking bones: all are perceived as painful, for good Darwinian reasons. Our artificially selected animals will enjoy having their skin pierced, will actively seek to break their own bones, will bask in a temperature so hot or so cold as to endanger their survival.

Similar artificial selection would work with humans. Not only could you breed for tastes, you could breed for callousness, sympathy, loyalty, slothfulness, piety, meanness or the Protestant work ethic. This is a less radical claim than it sounds, for genes don't fix behaviour deterministically, they only contribute quantitatively to statistical tendencies. Nor, as we saw when discussing the

---

* Marian Stamp Dawkins, the author of *Animal Suffering* and our leading investigator of the subject, has discussed with me the possibility that selective breeding of this sort might in theory provide a solution to some of the ethical problems of intensive animal husbandry. For instance, if present-day hens are unhappy in the confined conditions of battery cages, why not breed a race of hens that positively enjoy such conditions? She notes that people tend to greet such suggestions with repugnance (or humour in the case of Douglas Adams' brilliant *The Restaurant at the End of the Universe*, in which a large bovine quadruped approaches the table and announces itself as 'Your dish of the day', explaining that its kind has been bred to want to be eaten). Perhaps the idea conflicts with some deep-seated human value, possibly some version of what has been called the 'yuk factor'. It is hard to see that it falls foul of dispassionate utilitarian reasoning, provided we could be sure the selective breeding genuinely changed the animal's perception of pain, rather than – horrifying thought – changing its method of responding to pain while leaving the perception of pain intact.

values of science, does it imply a single gene for each of these complicated things, any more than the feasibility of breeding racehorses implies a single gene for speed. In the absence of artificial breeding, our own values are presumably influenced by natural selection under conditions that prevailed in the Pleistocene epoch in Africa.

Humans are unique in many ways. Perhaps our most obviously unique feature is language. Whereas eyes have evolved between forty and sixty times independently around the animal kingdom,[*] language has evolved only once.[†] It seems learned but there is strong

---

* It is in this spirit that I chose 'The fortyfold path to enlightenment' as my title for the chapter on the evolution of the eye in *Climbing Mount Improbable*. A whole chapter was necessary because, from William Paley on, the eye has been such a favourite of creationists seeking to apply what I called 'the Argument from Personal Incredulity'. Even Darwin confessed that the evolution of the eye seemed at first blush implausible. But his confession was a temporary rhetorical ploy, for he went on to show how easy it is to explain its gradual evolution. It is almost as though life is positively eager to evolve eyes, based on a variety of optical principles. Unlike language, which is the point I am making in this essay.

† This statement might be disputed, depending on the disputant's definition of language. Honey bees tell each other, with quantitative precision, how far away food is to be found, and in which direction relative to the sun. Vervet monkeys have three different 'words' for danger, depending on whether the threat is a snake, a bird or a leopard. I wouldn't call this language because it doesn't have the recursive, hierarchical embedding that gives human language its indefinite flexibility. Only humans can say things like: 'The leopard who has cubs, and who normally sits in the tree by the river in the direction of the mountain, is now crouching in the long grass beyond the hut belonging to the father of the chief.' Theoretically there is no limit to the depth of embedding of relative and prepositional clauses within one another, although keeping track of deep, multiple embedding makes demands on the brain's computational machinery. Steven Pinker's *The Language Instinct*

genetic supervision of the learning process. The particular language that we speak is learned, but the tendency to learn *language* rather than just any old thing is inherited and evolved specifically in our human line. We also inherit evolved rules for grammar. The exact readout of these rules varies from language to language, but their deep structure is laid down by the genes, and is presumably evolved by natural selection just as surely as our lusts and our bones. Evidence is good that the brain contains a language 'module', a computational mechanism that actively seeks to learn language and actively uses grammatical rules to structure it.

According to the young and thriving discipline of evolutionary psychology, the language learning module stands as exemplar of a whole set of inherited, special-purpose computational modules. We might expect modules for sex and reproduction, for analysing kinship (important for doling out altruism and avoiding dysgenic incest), for counting debts and policing obligations, for judging fairness and natural justice, perhaps for throwing projectiles accurately towards a distant target, and for classifying useful animals and plants. These modules will presumably be mediated by specific, built-in values.*

---

is a beautifully written, evolutionarily slanted introduction to such matters.

* The seminal book on evolutionary psychology, with chapters by many of its leading practitioners, is the volume edited by J. H. Barkow, L. Cosmides and J. Tooby, *The Adapted Mind*. Not long after this lecture was given, Steven Pinker's masterly *How the Mind Works* appeared. Evolutionary psychology, for reasons I don't understand, arouses incandescent hostility in quarters where I wouldn't expect it. The complaints seem to centre on particular studies that are poorly conceived or executed. But the existence of particular bad examples is no reason to dismiss an entire scientific discipline. The best practitioners of evolutionary psychology, Leda Cosmides, John Tooby, Steven Pinker, David Buss, Martin Daly, the late Margot Wilson and others, are good scientists by any standards.

If we turn Darwinian eyes on our modern, civilized selves and our predilections – our aesthetic values, our capacity for pleasure – it is important to wear sophisticated spectacles. Do not ask how a middle manager's ambitions for a bigger desk and a softer office carpet benefit his selfish genes. Ask, instead, how these urban partialities might stem from a mental module which was selected to do something else, in a different place and time. For office carpet, perhaps (and I mean *perhaps*) read soft and warm animal skins whose possession betokened hunting success. The whole art of applying Darwinian thinking to modern, domesticated humanity is to discern the correct rewriting rules. Take your question about the foibles of civilized, urban humanity and rewrite it back half a million years and out onto the African plains.

Evolutionary psychologists have coined the term *environment of evolutionary adaptedness*, or EEA, for that set of conditions in which our wild ancestors evolved. There's a lot that we don't know about the EEA; the fossil record is limited. Some of what we guess about it comes, through a kind of reverse engineering, from examining ourselves and trying to work out the kind of environment to which our attributes would have been well adapted.

We know that the EEA was located in Africa; probably, though not certainly, scrubby savannah. It is plausible that our ancestors lived in these conditions as hunter-gatherers, perhaps in something like the way modern hunter-gatherer tribes live in the Kalahari but, at least in earlier periods, with a less developed technology. We know that fire was tamed more than a million years ago by *Homo erectus*, the species that was probably our immediate predecessor in evolution. It is controversial when our ancestors dispersed out of Africa. We know that there were *Homo erectus* in Asia a million years ago, but many believe that nobody today is descended from those early migrants and that all surviving humans are the descendants of a second, more recent exodus of *Homo sapiens* out of Africa.*

---

* Current thinking favours several excursions out of Africa, and genetic evidence suggests a bottleneck, that is, a temporary,

Whenever the exodus, there has evidently been time for humans to adapt to non-African conditions. Arctic humans are different from tropical. We northerners have lost the black pigmentation that our African ancestors presumably had. There has been time for biochemistries to diverge in response to diet. Some peoples – perhaps those with herding traditions – retain into adulthood the ability to digest milk. In other peoples, only children can digest it; the adults suffer from the condition known as lactose intolerance. Presumably the differences have evolved by natural selection in different culturally determined environments. If natural selection has had time to shape our bodies and our biochemistry since some of us left Africa, it should also have had time to shape our brains and our values. So we needn't necessarily pay total heed to specifically African aspects of the EEA. Nevertheless, the genus *Homo* has spent at least nine-tenths of its time in Africa, and the hominins have spent 99 per cent of their time in Africa, so, insofar as our values are inherited from our ancestors, we might still expect a substantial African influence.

Various researchers, most notably Gordon Orians of the University of Washington, have examined aesthetic preferences for various kinds of landscapes. What kinds of environments do we

---

dramatic reduction in the population from which all non-Africans are descended, some time under one hundred thousand years ago. Yan Wong, in the second edition of *The Ancestor's Tale* which he co-authored with me, managed to use my genome (which happened to have been fully sequenced for a different purpose concerned with a television documentary) to estimate the population size at various times in the past. He did it by comparing my maternal genes and my paternal genes, estimating, for each pair, the time elapsed since they 'coalesced', i.e. since they split off from a common ancestral gene. A significant majority of my pairs of genes coalesced about sixty thousand years ago. This suggests that the population was briefly very small about sixty thousand years ago – hence a 'bottleneck'. It is probable that this bottleneck represents a particular out-of-Africa migration event.

seek to recreate in our gardens? These workers try to relate the sorts of places we find attractive to the sorts of places that our wild ancestors would have encountered as nomads wandering from camp site to camp site in the EEA. For example, we might be expected to like trees of the genus *Acacia* or other trees that resemble them. We might prefer landscapes in which the trees were low and dotted about, rather than deep forest landscapes, or deserts, both of which might carry threatening messages for us.

There seem to be some grounds for suspicion of this kind of work. Less justified would be a general scepticism that anything so complex or highflown as preference for a landscape could possibly be programmed into the genes. On the contrary, there is nothing intrinsically implausible about such values being inherited. Once again, a sexual parallel comes to mind. The act of sexual intercourse, if we contemplate it dispassionately, is pretty bizarre. The idea that there could be genes 'for' enjoying this preposterously unlikely act of rhythmic insertion and withdrawal might strike us as implausible in the extreme. But it is inescapable if we accept that sexual desire has evolved by Darwinian selection. Darwinian selection can't work if there are no genes to select. And if we can inherit genes for enjoying penile insertion, there is nothing inherently implausible in the idea that we could inherit genes for admiring certain landscapes, enjoying certain kinds of music, hating the taste of mangoes or anything else.

Fear of heights, manifesting itself in vertigo and in the common dreams of falling, might well be natural in species that spend a good deal of their time up trees, as our ancestors did. Fear of spiders, snakes and scorpions might with benefit be built into any African species. If you have a nightmare about snakes, it is just possible that you are dreaming not about symbolic phalluses but actually about *snakes*. Biologists have often noted that phobic reactions are commonly exhibited towards spiders and snakes, almost never to electric bulb sockets and motor cars. Yet in our temperate and urban world snakes and spiders no longer constitute a source of danger, while electric sockets and cars are potentially lethal.

It is notoriously hard to persuade drivers to slow down in fog, or refrain from tailgating at high speed. The economist Armen Alchian ingeniously suggested abolishing seatbelts and compulsorily fixing a sharp spear to all cars, pointing from the centre of the steering wheel straight at the driver's heart. I think I'd find it persuasive, whether for atavistic reasons I don't know. Also persuasive is the following calculation. If a car travelling at 80 miles per hour is abruptly slammed to a halt, this is equivalent to hitting the ground after falling from a tall building. In other words, when you are driving fast, it is as if you were hanging from the top of a high-rise building by a rope sufficiently thin that its probability of breaking is equal to the probability that the driver in front of you will do something really stupid. I know almost nobody who could happily sit on a windowsill up a skyscraper, few who unequivocally enjoy bungee-jumping. Yet almost everybody happily travels at high speed along motorways, even if they clearly understand in a cerebral way the danger they are in. I think it is quite plausible that we are genetically programmed to be afraid of heights and sharp points, but that we have to learn (and are not very good at) being afraid of travelling at high speeds.

Social habits that are universal among all peoples, such as laughing, smiling, weeping, religion, and a statistical tendency to avoid incest, are likely to have been present in our common ancestors too. Hans Hass and Irenäus Eibl-Eibesfeldt travelled the world clandestinely filming people's facial expressions, and concluded that there are cross-cultural universals in styles of flirting, of threatening, and in a fairly complicated repertoire of facial expressions. They filmed one child born blind, whose smile and other expressions of emotion were normal although she had never seen another face.

Children notoriously have a highly developed sense of natural justice, and 'not fair' is one of the first expressions to spring to the lips of a disgruntled child. This does not, of course, show that a sense of fairness is universally built in by the genes, but some might consider it suggestive in the same kind of way as the smile of the child born blind. It would be tidy if different cultures, the world over, shared the same ideas of natural justice. But there are some disconcerting differences. Most people attending this lecture would think it unjust to punish an individual for the crimes of his grandfather. Yet there are cultures where the transgenerational vendetta is taken for granted and is presumably regarded as naturally just.* This perhaps suggests that, at least in detail, our sense of natural justice is pretty flexible and variable.

Continuing our guesswork about the world of our ancestors, the EEA, there are reasons to think that they lived in stable bands, either roving and foraging like modern baboons, or perhaps more settled, in villages like present-day hunter-gatherers such as the Yąnomamö of the Amazon jungle. In either case, stability of grouping means that individuals would tend to encounter the same individuals repeatedly through their lives. Seen through Darwinian eyes, this could have had important consequences for the evolution of our values. In particular, it might help us to understand why, from the point of view of our selfish genes, we are so absurdly nice to each other.

It is not quite as absurd as it might naively appear. Genes may be selfish, but this is far from saying that individual organisms must be harsh and selfish. A large purpose of the doctrine of the selfish gene is to explain how selfishness at the gene level can lead to altruism at the level of the individual organism. But that only covers altruism as a kind of selfishness in disguise: first, altruism towards kin (nepotism); and second, boons given in the mathematical expectation of reciprocation (you help me and I'll repay you later).

This is where our assumption of life in villages or tribal bands can help – in two ways. First, there would probably have been a degree of inbreeding, as my colleague W. D. Hamilton has argued. Although, like many other mammals, humans go out of

* And sanctioned by the ultimate role model: ' . . . for I the LORD thy God am a jealous God, visiting the iniquity of the fathers upon the children unto the third and fourth generation of them that hate me' (Exodus 20: 5).

their way to combat the extremes of inbreeding, nevertheless neighbouring tribes frequently speak mutually unintelligible languages and practise incompatible religions, which inevitably limits crossbreeding. Assuming various low rates of between-village migration, Hamilton calculated the expected levels of genetic resemblance within tribes as compared with between tribes. His conclusion was that, under plausible assumptions, fellow village members might as well be brothers by comparison with outsiders from other villages.

Such conditions in the EEA would tend to favour xenophobia: 'Be unpleasant to strangers not from your own village, because strangers are statistically unlikely to share the same genes.' It is too simple to conclude that, conversely, natural selection in tribal villages would necessarily have favoured general altruism: 'Be nice to anyone you meet, because anyone you meet is statistically likely to share your genes for general altruism.'* But there could be

---

* I didn't have time to spell out in the lecture why it is too simple. The reason is that fellow villagers are not only likely to be your closest relatives; they are also your closest rivals for food, mates and other resources. For purposes of kin selection calculation, relatedness is computed not as an absolute number, but as the increment over and above a baseline of relatedness to random members of the population. In a close-knit, inbred village everybody you meet is likely to be a cousin. Kin selection theory predicts altruism towards individuals who are closer than the average, even when the average is already pretty close. In these circumstances, where a village consists of cousins, kin selection theory predicts xenophobia towards strangers from outside the village. My colleague Alan Grafen, in *Oxford Surveys in Evolutionary Biology* (1985), developed a beautiful geometric model, in my opinion by far the best way to explain the true meaning of r, the coefficient of relatedness which lies at the heart of kin selection theory. Many people who rely on popular accounts of Hamilton's theory are confused by the apparent mismatch between the values for r (0.5 for full siblings, 0.125 for first cousins) and the fact that all of us share more than 90 per cent of our genes

additional conditions in which this would indeed be so, and this was Hamilton's conclusion.

The other consequence of the village pattern follows from the theory of reciprocal altruism, which received a fillip in 1984 from the publication of Robert Axelrod's book *The Evolution of Cooperation*. Axelrod took the theory of games, specifically the game of prisoner's dilemma, and, abetted by Hamilton,* thought about it in an evolutionary way, using simple but ingenious computer models. His work has become well known and I shan't describe it in detail but summarize some relevant conclusions.

In an evolutionary world of fundamentally selfish entities, those individuals that cooperate turn out to be surprisingly likely to prosper. The cooperation is based not upon indiscriminate trust but upon swift identification and punishing of defection. Axelrod coined a measure, the 'shadow of the future', for how far ahead, on average, individuals can expect to continue meeting. If the shadow of the future is short, or if individual identification or its equivalent is hard, mutual trust is unlikely to develop and universal defection becomes the rule. If the shadow of the future is long, relationships of initial trust, tempered by suspicion of betrayal, are likely to develop. Such would have been the case in the EEA if our speculations about tribal villages or roving bands are correct. We therefore might expect to find, in ourselves, deep-seated tendencies towards what may be called 'suspicious trust'.

We should also expect to find in ourselves special-purpose

---

with each other. I give an example of this below in the article 'Twelve misunderstandings of kin selection' (page 163). Grafen's geometric model really gets across the point, in an intuitively vivid way, that *r* is the extra closeness over and above the baseline sharing among the whole population.

* In my foreword to the 2006 Penguin edition of *The Evolution of Cooperation*, I explained how I introduced Axelrod to Hamilton. I am quite proud to have instigated their fruitful collaboration, combining evolutionary with social science theory.

brain modules for calculating debt and repayment, for reckoning who owes how much to whom, for feeling pleased when one gains (but perhaps even more displeased when one loses), for mediating the sense of natural justice that I have already mentioned.

Axelrod went on to apply his version of game theory to the special case where individuals bear conspicuous labels. Suppose the population contains two types, arbitrarily called the reds and the greens. Axelrod concluded that, under plausible conditions, a strategy of the following form would be evolutionarily stable: 'If red, be nice to reds but nasty to greens; if green, be nice to greens but nasty to reds.' This follows regardless of the actual nature of redness and greenness, and regardless of whether the two types differ in any other respect at all. So, superimposed over the 'suspicious trust' I have mentioned, we should not be surprised to see discrimination of this kind.

What might 'red' versus 'green' correspond to in real life? Plausibly, own tribe versus other tribe. We have reached, via a different theory, the same conclusion as Hamilton with his inbreeding calculations. So the 'village model' leads us, by two quite distinct lines of theory, to expect ingroup altruism jockeying with tendencies to xenophobia.

Now, selfish genes are not conscious little agents, taking decisions for their own future good. The genes that survive are the ones that wired up ancestral brains with appropriate rules of thumb, actions that had the consequence, in ancestral environments, of assisting survival and reproduction. Our modern urban environment is very different, but the genes cannot be expected to have adjusted – there hasn't been time for the slow process of natural selection to catch up. So the same rules of thumb will be acted out as if nothing had happened. From the selfish genes' point of view it is a mistake, like our love of sugar in a modern world where sugar is no longer scarce and rots our teeth. It is entirely to be expected that there should be such mistakes. Perhaps, when you pity and help a beggar in the street, you are the misfiring instrument of a Darwinian rule of thumb set up in a tribal past when things were very different. I hasten to add that I use the word 'misfiring' in a strictly Darwinian sense, not as an expression of my own values.

So far so good, but there is probably more to goodness than that. Many of us seem generous beyond what would pay on 'selfishness in disguise' grounds, even assuming that we once lived in inbred bands who could expect a lifetime of opportunities for mutual repayment. If I live in such a world, I shall ultimately benefit if I build up a reputation for trustworthiness, for being the kind of person with whom you can enter into a bargain without fear of betrayal. As my colleague Matt Ridley puts it in his admirable book *The Origins of Virtue*, 'Now, suddenly, there is a new and powerful reason to be nice: to persuade people to play with you.' He quotes the economist Robert Frank's experimental evidence that people are good at rapidly sizing up, in a roomful of strangers, who can be trusted and who is likely to defect. But even that is still, in a sense, selfishness in disguise. The following suggestion may not be.

I think uniquely in the animal kingdom, we make good use of the priceless gift of foresight. Contrary to popular misunderstandings, natural selection has no foresight. It couldn't have, for DNA is just a molecule and molecules can't think. If they could, they'd have seen the danger presented by contraception and nipped it in the bud long ago. But brains are another matter. Brains, if they are big enough, can run all sorts of hypothetical scenarios through their imaginations and calculate the consequences of alternative courses of action. If I do such-and-such I'll gain in the short term. But if I do so-and-so, although I'll have to wait for my reward, it'll be bigger when it comes. Ordinary evolution by natural selection, though an immensely powerful force for technical improvement, cannot look ahead in this way.*

Our brains were endowed with the facility to set up goals and

---

* I'm fond of quoting Sydney Brenner, the eminent molecular geneticist, on the point. He satirically imagines a naive biologist speculating about a particular gene being favoured in the Cambrian era because 'it might come in handy in the Cretaceous' (hear it in a

purposes. Originally, these goals would have been strictly in the service of gene survival: proximally the goal of killing a buffalo, the goal of finding a new waterhole, the goal of kindling a fire, and so on. Still in the interests of gene survival, it was an advantage to make these goals as flexible as possible. New brain machinery, capable of deploying a hierarchy of reprogrammable subgoals within goals, started to evolve.

Imaginative forethought of this kind was originally useful but (in the genes'-eye view) it got out of hand. Brains as big as ours, as I've already argued, can actively rebel against the dictates of the naturally selected genes that built them. Using language, that other unique gift of the ballooning human brain, we can conspire together and devise political institutions, systems of law and justice, taxation, policing, public welfare, charity, care for the disadvantaged. We can invent our own values. Natural selection gives rise to these only at second remove, by making brains that grow big. From the point of view of the selfish genes our brains raced away with their emergent properties, and my personal value system regards this with a distinctly positive sign.

The tyranny of the texts

I have already disposed of one source of scepticism about my notion of rebellion against the selfish genes. Radical, left-wing scientists wrongly smelled a concealed Cartesian dualism. A different kind of scepticism comes from religious sources. Time and again, religious critics have said to me something like this. It's all very well issuing a call to arms against the tyranny of the selfish genes, but how do you decide what to put in its place? It's all very well sitting round a table with our big brains and our gift of foresight, but how are we going to agree on a set of values, how shall we decide what is good and what bad? What if somebody round the table advocates

---

sardonically witty South African accent and accompanied by a wicked twinkle in the eye).

cannibalism as the answer to the world's protein shortage, what ultimate authority can we call up to dissuade them? Aren't we going to be sitting in an ethical vacuum where, in the absence of strong, textual authority, anything goes? Even if you don't believe the existence claims of religion, don't we need religion as a source of ultimate values?

This is a genuinely difficult problem. I think we largely *are* in an ethical vacuum, and I mean all of us. If the hypothetical advocate of cannibalism was careful to specify road kills, who are already dead, he might even claim moral superiority over those who kill animals in order to eat them. There are still, of course, good counter-arguments; for instance, the 'distress to relatives' argument applies more strongly to humans than to other species; or there's the slippery slope argument ('If we get used to eating human road kills, it will be just a short step to . . .' and so on).

So, I am not minimizing the difficulties. But what I will now say – and it's putting it mildly – is that we are no *worse* off than we were when we relied on ancient texts. The moral vacuum we now feel ourselves to be in has always been there, even if we haven't recognized it. Religious people are already entirely accustomed to picking and choosing *which* texts from holy books they obey and which they reject. There are passages of the Judeo-Christian Bible which no modern Christian or Jew would wish to follow. The story of Isaac's narrowly averted sacrifice by his father Abraham strikes us moderns as a shocking piece of child abuse, whether we read it literally or symbolically.

Jehovah's appetite for the smell of burning flesh has no appeal for modern tastes. In Judges, chapter 11, Jephthah made a vow to God that, if God could guarantee Jephthah's victory over the Children of Ammon, Jephthah would, without fail, sacrifice as a burnt offering 'whatsoever cometh forth of the doors of my house to meet me, when I return'. As luck would have it, this turned out to be Jephthah's own daughter, his only child. Understandably he rent his clothes, but there was nothing he could do about it, and his daughter very decently agreed that she should be sacrificed. She asked only that she should be allowed to go into the mountains for two months to bewail her virginity. At the end of this time, Jephthah slaughtered his own daughter and turned her into a burnt offering as Abraham nearly had his son. God was not moved to intervene on this occasion.

Much of what we read of Jehovah makes it hard to see him as a good role model, whether we think of him as a factual or fictitious character. The texts show him to be jealous, vindictive, spiteful, capricious, humourless and cruel.* He was also, in modern terms, sexist, and an inciter to racial violence. When Joshua 'utterly destroyed all that was in the city both man and woman, young and old, and ox, and sheep, and ass, with the edge of the sword', you might ask what the citizens of Jericho had done to deserve such a terrible fate. The answer is embarrassingly straightforward: they belonged to the wrong tribe. God had promised some *Lebensraum* to the Children of Israel, and the indigenous population was in the way.

> But of the cities of these people, which the LORD thy God doth give thee for an inheritance, thou shalt save alive nothing that breatheth;
>> But thou shalt utterly destroy them; namely, the Hittites, and the Amorites, the Canaanites, and the Perizzites, the Hivites, and the Jebusites; as the LORD thy God hath commanded thee.[†]

---

* An expanded version of this list of unpleasant adjectives formed the opening paragraph of chapter 2 of *The God Delusion*, where it has become somewhat infamous for giving 'offence'. Every one of those adjectives can be justified from scripture, as was demonstrated by my colleague Dan Barker. His splendid book, *God: the most unpleasant character in all fiction*, takes each one of my distasteful adjectives in order and documents them meticulously with quotations from the Bible – which he knows very well, as a sometime preacher who has now seen the light.

† Deuteronomy 20: 16–17. It's been suggested to me that my use of the German *Lebensraum* is, in the circumstances, offensive or 'inappropriate' (to use the cant word). But I find it impossible to think of any word that is more spot-on, nail-on-head appropriate.

Now, of course, I'm being terribly unfair. The one thing a historian must never do is judge one era by the standards of a later era. But that is precisely my *point*. You cannot have it both ways. If you claim the right to pick and choose the nice bits of the Bible and sweep the nasty bits under the carpet, you have sold the pass. You have admitted that you do not, as a matter of fact, get your values from an ancient and authoritative holy book. You are demonstrably getting your values from some modern source, some contemporary liberal consensus or whatever it is. Otherwise by what criterion do you choose the good bits of the Bible while rejecting, say, Deuteronomy's clear injunction to stone non-virgin brides to death?

Wherever this contemporary liberal consensus may come from, I am entitled to appeal to it when I explicitly reject the authority of my ancient text – the DNA – just as you are entitled to appeal to it when you implicitly reject your – rather less ancient – texts from human scriptures. We can all sit down together and work out the values we want to follow. Whether we are talking about four-thousand-year-old parchment scrolls, or four-thousand-million-year-old DNA, we are all entitled to throw off the tyranny of the texts.

AFTERWORD

Although the onus is not on me to say where religious people find the modern consensus whereby they decide which are the good verses of the Bible and which the horrible ones, there is nevertheless a genuinely interesting question lurking here. Where do our twenty-first-century values come from, as opposed to the relatively nasty ones of earlier centuries? What has changed, such that in the 1920s 'votes for women' was a daringly radical proposal, leading to riots in the streets, whereas now to forbid women the vote is regarded as an obvious outrage? Looking back to earlier centuries, Steven Pinker's *The Better Angels of our Nature* and Michael Shermer's *The Moral Arc* document inexorable improvements in our values. Improvements by whose standards? By the standards of modern times, of course – a line of reasoning which, although circular, is not viciously so.

Think of the slave trade, think of killing as a spectator sport in the Roman Colosseum; of bear-baiting, burning at the stake, treatment of prisoners including prisoners of war before the Geneva Convention. Think of warfare itself, and consider the wholesale and deliberate bombing of cities in the 1940s against the fact that modern air forces feel the need to apologize when civilian targets are accidentally hit. The moral arc shows some erratic zigzagging, but the trend is unmistakably in one direction. Whatever caused the change, it wasn't religion. But what was it?

'Something in the air'? That sounds mystical but it can be rendered in sensible terms. I liken the process to Moore's Law, which states that computer power has increased over the decades at a lawful rate, though nobody really knows why. Well, we understand it in a general way, but we don't know why it is so beautifully lawful (a straight line when plotted on a logarithmic scale). For some reason improvements in hardware and software, themselves the summed effects of lots of different kinds of detailed improvements, in different companies in different parts of the world, come together to yield Moore's Law. What are the equivalent trends that sum up to make the Shifting Moral *Zeitgeist*, with its overall unidirectional (albeit slightly more erratic) line? Again, the onus is not on me to name them, but I would guess it's some combination of the following:

legal decisions in courts of law;

speeches and votes in parliaments and houses of congress;

lectures, papers and books by moral and legal philosophers;

journalistic articles and newspaper editorials;

everyday conversations at dinner parties and in pubs, on radio and television

This all leads to an obvious next question. Whither the moral arc in future decades and centuries? Can you think of something that we accept with equanimity in 2017 but future centuries will regard with the same revulsion as we, today, view the slave trade or the railway wagons bound for Belsen and Buchenwald? I don't think it requires much imagination to think of at least one candidate. Don't the Belsen-bound railway wagons come, unwelcome, to mind when you drive behind one of those closed-in trucks with bewildered, fearful eyes peering through the ventilation slats?

Speaking up for science: an open letter to Prince Charles

YOUR ROYAL HIGHNESS,
Your Reith Lecture* saddened me. I have deep sympathy for your aims, and admiration for your sincerity. But your hostility to science will not serve those aims; and your embracing of an ill-assorted jumble of mutually contradictory alternatives will lose you the respect that I think you deserve. I forget who it was† who remarked: 'Of course we must be open-minded, but not so open-minded that our brains drop out.'

Let's look at some of the alternative philosophies which you seem to prefer over scientific reason. First, intuition, the heart's wisdom 'rustling like a breeze through the leaves'. Unfortunately, it depends whose intuition you choose. Where aims (if not methods) are concerned, your own intuitions coincide with mine. I

---

* The annual Reith Lectures, originally broadcast on radio, now also on television, are sponsored by the BBC to commemorate its founding Director General, Lord Reith, an austere Scot whose high ideals the BBC has largely forsaken. It is still considered a great honour in Britain to be invited to give the Reith Lectures. Unusually the 2000 series, on 'Respect for the Earth', was divided among five lecturers, of whom Prince Charles was one. This open letter replying to it was first published in the *Observer* on 21 May 2000.

† It's often attributed to me but, much as I'd like to own it, I'm pretty sure I got it from somewhere else.

67

wholeheartedly share your aim of long-term stewardship of our planet, with its diverse and complex biosphere.*

But what about the instinctive wisdom in Saddam Hussein's black heart?† What price the Wagnerian wind that rustled Hitler's

---

* The Prince's concerns have become more urgent in the years since he gave his lecture. The signs of drastic climate change have become ever more unmistakable, and there is now serious talk that we may have passed a point of no return. Meanwhile, the incoming US President has publicly announced his view that climate change is a 'Chinese hoax'. It is still (just) possible to entertain a (decreasingly plausible) argument that humans are not responsible for such trends as the disappearing polar ice. But the reality of dangerous and worsening climate change itself is now plain for all but the deluded to see. In the face of this looming catastrophe, including worldwide flooding of low-lying areas, it is all the more important not to cry wolf with respect to lesser problems in the way that, unfortunately, Prince Charles is wont to do.

† I went on record to deplore, at the time, the execution of Saddam Hussein, not simply because of general opposition to the death penalty but for scientific reasons. I would also have spared Hitler's life if he hadn't taken it himself. We need all the information we can get to understand the mentality of such monsters; and – because sociopaths are not all that rare – to understand how exceptional examples like Hitler manage to acquire and retain power over other people and even win elections. Was Hitler really a mesmerizing orator with hypnotically compelling eyes, as alleged by some who knew him? Or was that an illusion fostered in hindsight by the aura of power? How would a jailed Hitler have responded to alternative approaches to making him see reason, for example to quiet and sober arguments questioning his pathological hatred of Jews? Could we have gained an understanding of powerful psychopathology which might have been useful for the future? Was there something in Hitler's childhood, or Saddam Hussein's, or their early education, that set them on the path to their adult selves? Could some kind of educational reform forestall similar horrors in the future? Killing such odious specimens might

twisted leaves? The Yorkshire Ripper heard religious voices in his head urging him to kill. How do we decide *which* intuitive inner voices to heed?

This, it is important to say, is not a dilemma that science can solve. My own passionate concern for world stewardship is as emotional as yours. But where I allow feelings to influence my aims, when it comes to deciding the best method of achieving them I'd rather think than feel. And thinking, here, means scientific thinking. No more effective method exists. If it did, science would incorporate it.

Next, Sir, I think you may have an exaggerated idea of the naturalness of 'traditional' or 'organic' agriculture. Agriculture has always been unnatural. Our species began to depart from our natural hunter-gatherer lifestyle as recently as ten thousand years ago – too short a period to measure on the evolutionary timescale.

Wheat, be it ever so wholemeal and stoneground, is not a natural food for *Homo sapiens*. Nor is milk, except for children. Almost every morsel of our food is genetically modified – admittedly mostly by artificial selection not artificial mutation, but the end result is the same. A wheat grain is a genetically modified grass seed, just as a pekinese is a genetically modified wolf. Playing God? We've been playing God for centuries!

The large, anonymous crowds in which we now teem began with the agricultural revolution, and without agriculture we could survive in only a tiny fraction of our current numbers. Our high population is an agricultural (and technological and medical) artefact. It is *far* more unnatural than the population-limiting methods condemned as unnatural by the Pope. Like it or not, we are stuck with agriculture, and agriculture – *all* agriculture – is unnatural. We sold that pass ten thousand years ago.

Does that mean there's nothing to choose between different kinds of agriculture when it comes to sustainable planetary welfare?

---

satisfy primal vengeance, but closes off avenues of research that could help avoid recurrences.

Certainly not. Some are much more damaging than others, but it's no use appealing to 'nature', or to 'instinct', in order to decide which ones. You have to study the evidence, soberly and reasonably – scientifically. Slashing and burning (incidentally, no agricultural system is closer to being 'traditional') destroys our ancient forests. Overgrazing (again, widely practised by 'traditional' cultures) causes soil erosion and turns fertile pasture into desert. Moving to our own modern tribe, monoculture, fed by powdered fertilizers and poisons, is bad for the future; indiscriminate use of antibiotics to promote livestock growth is worse.

Incidentally, one worrying aspect of the hysterical opposition to the *possible* risks from GM crops is that it diverts attention from *definite* dangers which are already well understood but largely ignored. The evolution of antibiotic-resistant strains of bacteria is something that a Darwinian might have foreseen from the day antibiotics were discovered. Unfortunately the warning voices have been rather quiet, and now they are drowned by the baying cacophony: 'GM GM GM GM GM GM!'

Moreover if, as I expect, the dire prophecies of GM doom fail to materialize, the feeling of let-down may spill over into complacency about real risks. Has it occurred to you that our present GM brouhaha may be a terrible case of crying wolf?

Even if agriculture could be natural, and even if we could develop some sort of instinctive rapport with the ways of nature, would nature be a good role model? Here, we must think carefully. There really is a sense in which ecosystems are balanced and harmonious, with some of their constituent species becoming mutually dependent. This is one reason the corporate thuggery that is destroying the rainforests is so criminal.

On the other hand, we must beware of a very common misunderstanding of Darwinism. Tennyson was writing before Darwin but he got it right. Nature really is red in tooth and claw. Much as we might like to believe otherwise, natural selection, working within each species, does not favour long-term stewardship. It favours short-term gain. Loggers, whalers and other profiteers who squander the future for present greed are only doing what wild creatures have done for three billion years.

No wonder T. H. Huxley, Darwin's bulldog, founded his ethics on a repudiation of Darwinism. Not a repudiation of Darwinism as science, of course, for you cannot repudiate truth. But the very fact that Darwinism is true makes it even more important for us to fight against the naturally selfish and exploitative tendencies of nature. We can do it. Probably no other species of animal or plant can. We can do it because our brains (admittedly given to us by natural selection for reasons of short-term Darwinian gain) are big enough to see into the future and plot long-term consequences. Natural selection is like a robot that can only climb uphill, even if this leaves it stuck on top of a measly hillock. There is no mechanism for going downhill, for crossing the valley to the lower slopes of the high mountain on the other side. There is no natural foresight, no mechanism for warning that present selfish gains are leading to species extinction – and indeed, 99 per cent of all species that have ever lived are extinct.

The human brain, probably uniquely in the whole of evolutionary history, can see across the valley and can plot a course away from extinction and towards distant uplands. Long-term planning – and hence the very possibility of stewardship – is something utterly new on the planet, even alien. It exists only in human brains. The future is a new invention in evolution. It is precious. And fragile. We must use all our scientific artifice to protect it.

It may sound paradoxical, but if we want to sustain the planet into the future, the first thing we must do is stop taking advice from nature. Nature is a short-term Darwinian profiteer. Darwin himself said it: 'What a book a devil's chaplain* might write on the clumsy, wasteful, blundering, low, and horridly cruel works of nature.'

Of course that's bleak, but there's no law saying the truth has to

* I borrowed Darwin's phrase as the title of my previous anthology, published in 2003.

be cheerful; no point shooting the messenger – science – and no sense in preferring an alternative world-view just because it feels more comfortable. In any case, science isn't all bleak. Nor, by the way, is science an arrogant know-all. Any scientist worthy of the name will warm to your quotation from Socrates: 'Wisdom is knowing that you don't know.' What else drives us to find out?

What saddens me most, Sir, is how much you will be missing if you turn your back on science. I have tried to write about the poetic wonder of science myself,* but may I take the liberty of presenting you with a book by another author? It is *The Demon-Haunted World* by the lamented Carl Sagan. I'd call your attention especially to the subtitle: *Science as a Candle in the Dark.*

AFTERWORD

One important principle which I should have mentioned by name in my letter to Prince Charles is the precautionary principle. He is certainly right that, where new and untried technologies are concerned, we should lean towards the conservative. If something is untried and we don't know the consequences, it behoves us to err on the side of caution, especially where long-term futures are at stake. It is the precautionary principle that requires apparently promising new cancer drugs to jump through hoops and over hurdles before being certified for general use. Such risk-averse hurdles can reach ludicrous heights, as when patients who are already at death's door are denied access to experimental drugs which just might save their lives but which have yet to be certified as 'safe'. Terminal patients have a different conception of 'safe'. But in general it is hard to deny the wisdom of the precautionary principle, sensibly balanced against the huge advantages that scientific innovation can bring.

While I'm on the precautionary principle, please forgive a digression into contemporary politics. Normally I would fight shy

* In *Unweaving the Rainbow.*

of up-to-the-minute currency, for fear of anachronizing future editions of a book. J. B. S. Haldane's and Lancelot Hogben's otherwise admirable writings of the 1930s are marred by political barbs that are obtrusively incomprehensible today. Unfortunately, the repercussions of at least two of the political events of 2016 – Britain's vote to leave the European Union and America's repudiation of international agreements on climate change – have little chance of limitation to the short term. So, without apology, I speak of 2016 politics.

In 2016 our then Prime Minister David Cameron caved in to pressure from his backbenchers to hold a referendum on British membership of the EU. This was a question of immense complexity involving sophisticated economic ramifications, the full extent of which became only too apparent later in the year when prodigal regiments of lawyers and civil servants had to be employed to cope with the administrative and legal load. If ever there was a matter for lengthy parliamentary debate and cabinet discussion heavily informed by advice from highly qualified experts, it was membership of the EU. Could there be a question *less* suited to a single plebiscite decision? And yet we were told to mistrust experts ('You, the voter, are the expert here') by politicians who presumably would demand an expert surgeon to remove their appendix or an expert pilot to fly their plane. So the decision was handed over to non-experts like me, even people whose stated motives for voting included 'Well, it's nice to have a change,' and 'Well, I preferred the old blue passport to the European purple one.' For the sake of short-term political manoeuvring within his own party, David Cameron played Russian roulette with the long-term future of his country, of Europe, even of the world.

And so, to the precautionary principle. The referendum was about a major change, a political revolution whose pervasive effects would persist for decades if not longer. A huge constitutional change, the sort of change where, if ever, the precautionary principle should have been paramount. When it comes to constitutional amendments, the United States requires a two-thirds majority in both houses of Congress followed by ratification by three-quarters of the state legislatures. Arguably that bar is set a bit too high, but the principle is sound. David Cameron's referendum, by contrast,

THE VALUE(S) OF SCIENCE

asked for only a simple majority on a single yes/no question. Did it not occur to him that so radical a constitutional step might merit stipulation of a two-thirds majority? Or at least 60 per cent? Perhaps a minimum voter turnout to make sure such a major decision was not taken by a minority of the electorate? Maybe a second vote, a fortnight later, to make sure the populace really meant it? Or a second round a year later, when the terms and consequences of withdrawal had become at least minimally apparent? But no, all Cameron demanded was anything over 50 per cent in a single yes/no vote, at a time when opinion polls were yo-yo-ing up and down and the likely result was changing day by day. It is said that a leftover statute of British common law stipulates that 'no idiot shall be admitted to parliament'. You'd think at least the stricture might apply to Prime Ministers.

As with Prince Charles's hostility to aspects of scientific food production, the precautionary principle should be applied judiciously. It can go too far, and it's arguable, as I said before, that for constitutional amendments in the United States the bar is set too high. It is widely agreed that the Electoral College is an undemocratic anachronism, but also widely accepted that it is almost impossible to abolish it because of the high hurdle of constitutional amendment. It seems that where huge decisions with far-reaching implications, such as constitutional amendments, are concerned, observance of the precautionary principle in politics needs to be pitched somewhere between its current positions in the too risk-averse United States, where the written constitution has fossilized into an object of near-sacred veneration, and in Britain, whose unwritten constitution leaves the door open to the kind of reckless irresponsibility of Cameron's EU referendum.

Finally, since this disquisition on the precautionary principle comes at the end of a letter to the Heir Apparent, what of that historic plank of our British unwritten constitution, the hereditary monarchy itself? The monarch is also, of course, head of the Church of England. Her many titles include 'Defender of the Faith' which, make no mistake, means specifically defender of one religion against a rival religion or denomination. When the title was invented, the possibility that an heir might grow up to be an atheist (as seems more than likely if current trends continue), or might have a Muslim stepfather (as nearly happened in living memory) never entered anyone's head.

Though stripped of most of the dictatorial powers of her earlier predecessors, the monarch still has advisory powers (and Elizabeth II is richly experienced in using them, having gone through no fewer than fourteen Prime Ministers). In extreme cases, the monarch is constitutionally able to dissolve Parliament on her sole initiative, although to do so would precipitate a crisis of uncertain and hazardous outcome. Even setting aside this unlikely possibility, many people find the idea of a hereditary monarchy hard to justify, and there are some who advocate a respectful termination of the institution on the death of the present Queen – which I, for one, hope will be a long time in the future.

Whenever I talk to avid British republicans, I cannot help making at least a glancing allusion to the precautionary principle. In various forms the monarchy has been soldiering on for well over a thousand years. What are you going to put in its place? A Facebook vote for head of state? King Becks and Queen Posh aboard the Royal Yacht *Boaty McBoatface*? There are, no doubt, better alternatives than my shamelessly elitist satire. There was a time when I would have pointed to the United States as a role model. But that was before 2016 showed us what the noble democratic ideal, when it turns sour, is capable of delivering.

Science and sensibility

WITH TREPIDATION AND humility, I find myself the only scientist in this list of lecturers.* Does it really fall to me alone to 'sound the century' for science; to reflect on the science that we bequeath to our heirs? The twentieth could be science's golden century: the age of Einstein, Hawking and relativity; of Planck, Heisenberg and quantum theory; of Watson, Crick, Sanger and molecular biology; of Turing, von Neumann and the computer; of Wiener, Shannon and cybernetics; of plate tectonics and radioactive dating of the rocks; of Hubble's Red Shift and the Hubble Telescope; of Fleming, Florey and penicillin; of moon landings; and – let's not duck the issue – of the hydrogen bomb. As George Steiner has noted, more scientists are working today than in all other centuries combined. Though also – to put that figure into alarming perspective – more people are alive today than have died since the dawn of recorded history.

Of the dictionary meanings of sensibility, I intend 'discernment,

* At the end of the twentieth century the BBC put on a series of lectures, broadcast on Radio 3, on the theme of 'Sounding the Century: what will the twentieth century leave to its heirs?' My contribution was delivered on 24 March 1998; other speakers included Gore Vidal, Camille Paglia and George Steiner. I was uncomfortably conscious of being the only scientist on the list, hence my opening sentences. Parts of the lecture found their way into *Unweaving the Rainbow*, which I was writing around the same time.

awareness' and 'the capacity for responding to aesthetic stimuli'. One might have hoped that, by century's end, science would have been incorporated into our culture, and our aesthetic sense have risen to meet the poetry of science. Without reviving the midcentury pessimism of C. P. Snow, I reluctantly find that, with only two years to run, these hopes are not realized. Science provokes more hostility than ever, sometimes with good reason, often from people who know nothing about it and use their hostility as an excuse not to learn. Depressingly many people still fall for the discredited cliché that scientific explanation corrodes poetic sensibility. Astrology books outsell astronomy. Television beats a path to the door of second-rate conjurors masquerading as psychics and clairvoyants. Cult leaders mine the millennium and find rich seams of gullibility: Heaven's Gate, Waco, poison gas in the Tokyo underground. The biggest difference from the last millennium is that folk Christianity has been joined by folk science fiction.

It should have been so different. The previous millennium, there was some excuse. In 1066, if only with hindsight, Halley's Comet could forebode Hastings, sealing Harold's fate and Duke William's victory. Comet Hale-Bopp in 1997 should have been different. Why do we feel gratitude when a newspaper astrologer reassures his readers that Hale-Bopp was not *directly* responsible for Princess Diana's death? And what is going on when thirty-nine people, driven by a theology compounded of *Star Trek* and the Book of Revelation, commit collective suicide, neatly dressed and with overnight bags packed by their sides, because they all believed that Hale-Bopp was accompanied by a spaceship come to 'raise them to a new plane of existence'? Incidentally, the same Heaven's Gate commune had ordered an astronomical telescope to look at Hale-Bopp. They sent it back when it came, because it was obviously defective: it failed to show the accompanying spaceship.

Hijacking by pseudoscience and bad science fiction is a threat to our legitimate sense of wonder. Hostility from academics sophisticated in fashionable disciplines is another, and I shall return to this. Populist 'dumbing down' is a third. The 'public understanding of science' movement, provoked in America by Sputnik and driven in Britain by alarm over a decline in science applicants at universities, is going demotic. A spate of 'Science Fortnights' and the like betrays a desperate anxiety among scientists to be loved. Whacky 'personalities', with funny hats and larky voices, perform explosions and funky tricks to show that science is fun, fun, fun.

I recently attended a briefing session urging scientists to put on 'events' in shopping malls, designed to lure people into the joys of science. We were advised to do nothing that might conceivably be a 'turn-off'. Always make your science 'relevant' to ordinary people – to what goes on in their own kitchen or bathroom. If possible, choose experimental materials that your audience can eat at the end. At the last event organized by the speaker himself, the scientific feat that really grabbed attention was the urinal which automatically flushed as soon as you stepped away. The very word science is best avoided, because 'ordinary people' find it threatening.*

When I protest, I am rebuked for my 'elitism'. A terrible word, but maybe not such a terrible thing? There's a great difference between an exclusive snobbery, which no one should condone, and a striving to help people raise their game and swell the elite. A calculated dumbing down is the worst, condescending and patronizing. When I said this in a recent lecture in the United States, a questioner at the end, no doubt with a warm glow in his white male heart, had the remarkable cheek to suggest that dumbing down might be especially necessary to bring 'minorities and women' to science.

I worry that to promote science as all larky and easy is to store

---

* I'm sceptical about the very notion of 'ordinary people'. The great Francis Crick was once persuaded by a publisher to write a book 'for ordinary people'. Understandably nonplussed by the commission, he was heard calling out to his colleague the distinguished neurologist V. S. Ramachandran, 'I say, Rama, do you know any ordinary people?'

up trouble for the future. Recruiting advertisements for the army don't promise a picnic, for the same reason. Real science can be hard but, like classical literature or playing the violin, worth the struggle. If children are lured into science, or any other worthwhile occupation, by the promise of easy frolics, what happens when they finally confront the reality? 'Fun' sends the wrong signals and might attract recruits for the wrong reasons.

Literary studies are at risk of becoming similarly undermined. Idle students are seduced into a debased 'Cultural Studies', where they will spend their time 'deconstructing' soap operas, tabloid princesses and tellytubbies. Science, like proper literary studies, can be hard and challenging but science is – again like proper literary studies – wonderful. Science is also useful; but useful is not all it is. Science can pay its way but, like great art, it shouldn't have to. And we shouldn't need whacky personalities and explosions to persuade us of the value of a life spent finding out why we have life in the first place.

Perhaps I'm being too negative, but there are times when a pendulum has swung too far and needs a push in the other direction. Certainly, practical demonstrations can make ideas vivid and preserve them in the mind. From Michael Faraday's Royal Institution Christmas Lectures to Richard Gregory's Bristol Exploratory, children have been excited by hands-on experience of true science. I was myself honoured to give the Christmas Lectures, in their modern televised form, with plenty of hands-on demonstrations. Faraday never dumbed down. I am attacking only the kind of populist whoring that defiles the wonder of science.

Annually in London there is a large dinner, at which prizes for the year's best science books are presented. One prize is for children's science books, and it recently went to a book about insects and other so-called 'ugly bugs'. Such language is not best calculated to arouse the poetic sense of wonder, but let that pass. Harder to forgive were the antics of the chair of the judges, a well-known television personality (who had credentials to present real science, before she sold out to 'paranormal' television). Squeaking with game-show levity, she incited the audience to join her in repeated choruses of audible grimaces at the contemplation of the horrible 'ugly bugs'. 'Eeeuurrrgh! Yuck! Yeeyuck! Eeeeeuurrrgh!' That kind of vulgarity demeans the wonder of science, and risks 'turning off' the very people best qualified to appreciate it and inspire others: real poets and true scholars of literature.

The true poetry of science, especially twentieth-century science, led the late Carl Sagan to ask the following acute question.

> How is it that hardly any major religion has looked at science and concluded, 'This is better than we thought! The Universe is much bigger than our prophets said, grander, more subtle, more elegant'? Instead they say, 'No, no, no! My god is a little god, and I want him to stay that way.' A religion, old or new, that stressed the magnificence of the Universe as revealed by modern science might be able to draw forth reserves of reverence and awe hardly tapped by the conventional faiths.

Given a hundred clones of Carl Sagan, we might have some hope for the next century. Meanwhile, in its closing years, the twentieth must be rated a disappointment as far as public understanding of science is concerned, while being a spectacular and unprecedented success with respect to scientific achievements themselves.*

What if we let our sensibility play over the whole of twentieth-century science? Is it possible to pick out a theme, a scientific leitmotif? My best candidate comes nowhere near doing justice to the richness on offer. The twentieth is the Digital Century. Digital discontinuity pervades the engineering of our time, but there is a sense in which it spills over into the biology and perhaps even the physics of our century.

The opposite of digital is analogue. When the Spanish Armada

---

* I was, perhaps, being unduly pessimistic there. I'm always, and was in the twentieth century too, encouraged by the large and enthusiastic audiences for science writers at festivals such as Hay and Cheltenham, and colleagues such as Steve Jones and Steven Pinker say the same.

was expected, a signalling system was devised to spread the news across southern England. Bonfires were set on a chain of hilltops. When any coastal observer spotted the Armada he was to light his fire. It would be seen by neighbouring observers, their fires would be lit, and a wave of beacons would spread the news at great speed far across the coastal counties.

How could we adapt the bonfire telegraph to convey more information? Not just 'The Spanish are here' but, say, the size of their fleet? Here's one way. Make your bonfire's size proportional to the size of the fleet. This is an analogue code. Clearly, inaccuracies would be cumulative. So, by the time the message reached the other side of the kingdom, the information about fleet size would have degraded to nothing. This is a general problem with analogue codes.

But now here's a simple digital code. Never mind the size of the fire, just build any serviceable blaze and place a large screen around it. Lift the screen and lower it again, to send the next hill a discrete flash. Repeat the flash a particular number of times, then lower the screen for a period of darkness. Repeat. The number of flashes per burst should be made proportional to the size of the fleet.

This digital code has huge virtues over the previous analogue code. If a hilltop observer sees eight flashes, eight flashes is what he passes along to the next hill in the chain. The message has a good chance of spreading from Plymouth to Dover without serious degradation. The superior power of digital codes has been clearly understood only in the twentieth century.

Nerve cells are like Armada beacons. They 'fire'. What travels along a nerve fibre is not electric current. It's more like a trail of gunpowder laid along the ground. Ignite one end with a spark, and the fire fizzes along to the other end.

We've long known that nerve fibres don't use purely analogue codes. Theoretical calculations show that they couldn't. Instead, they do something more like my flashing Armada beacons. Nerve impulses are trains of voltage spikes, repeated as in a machine gun. The difference between a strong message and a weak is not conveyed by the height of the spikes – that would be an analogue code and the message would be distorted out of existence. It is conveyed by the pattern of spikes, especially the firing rate of the machine gun. When you see yellow or hear middle C, when you smell turpentine or touch satin, when you feel hot or cold, the differences are being rendered, somewhere in your nervous system, by different rates of machine-gun pulses. The brain, if we could listen in, would sound like Passchendaele. In our meaning, it is digital. In a fuller sense it is still partly analogue: rate of firing is a continuously varying quantity. Fully digital codes, like Morse, or computer codes, where pulse patterns form a discrete alphabet, are even more reliable.

If nerves carry information about the world as it is now, genes are a coded description of the distant past. This insight follows from the 'selfish gene' view of evolution.

Living organisms are beautifully built to survive and reproduce in their environments. Or that is what Darwinians say. But actually it isn't quite right. They are beautifully built for survival in their ancestors' environments. It is because their ancestors survived – long enough to pass on their DNA – that our modern animals are well built. For they inherit the very same successful DNA. The genes that survive down the generations add up, in effect, to a description of what it took to survive back then. And that is tantamount to saying that modern DNA is a coded description of the environments in which ancestors survived. A survival manual is handed down the generations. A Genetic Book of the Dead.*

Like the longest chain of beacon fires, the generations are

* This phrase became the title of a chaper in *Unweaving the Rainbow* which developed the theme more fully. I argued that a well-informed biologist of the future, when presented with an animal – or with its DNA – should be able to 'read' the animal and reconstruct the environment in which its ancestors survived and reproduced. Not just the physical environment – weather, soil chemistry and so on – but the biological environment, the predators or prey, parasites or hosts, with which its ancestral lineage ran in evolutionary 'arms races'.

uncountably many. No surprise, then, that genes are digital. Theoretically the ancient book of DNA could have been analogue. But, for the same reason as for our analogue Armada beacons, any ancient book copied and recopied in analogue language would degrade to meaninglessness in very few scribe generations. Fortunately, human writing is digital, at least in the sense we care about here. And the same is true of the DNA books of ancestral wisdom that we carry around inside us. Genes are digital, and in the full sense not shared by nerves.

Digital genetics was discovered in the nineteenth century, but Gregor Mendel was ahead of his time and ignored. The only serious error in Darwin's world-view derived from the conventional wisdom of his age, that inheritance was 'blending' – analogue genetics. It was dimly realized in Darwin's time that analogue genetics was incompatible with his whole theory of natural selection. Even less clearly realized, it was also incompatible with obvious facts of inheritance.* The solution had to wait for the twentieth century, especially the neo-Darwinian synthesis of Ronald Fisher and others in the 1930s. The essential difference between classical Darwinism (which we now understand could not

---

* In 1867, the Scottish engineer Fleeming Jenkin pointed out that blending inheritance would remove variation from the population, generation by generation. By analogy, if you mix black paint with white you get grey, and no amount of mixing grey with grey will restore the original black and white. Therefore natural selection will rapidly find that it has no variation to choose from, therefore Darwin must be wrong. What Jenkin overlooked is that it is manifestly *false* that every generation is, as a matter of fact, greyer than its parents. He thought he was arguing against Darwin. He was actually arguing against manifest fact. Variation clearly does not dwindle as the generations go by. Did he but know it, far from disproving Darwin, Jenkin was actually disproving blending inheritance. He could have worked out Mendel's laws intuitively from the depths of an armchair, without bothering to grow peas in a monastery garden.

have worked) and neo-Darwinism (which does) is that digital genetics has replaced analogue.

But when it comes to digital genetics, Fisher and his colleagues of the synthesis didn't know the half of it. Watson and Crick opened floodgates to what has been, by any standards, a spectacular intellectual revolution – even if Peter Medawar was going too far when he wrote, in his 1968 review of Watson's *The Double Helix*, 'It is simply not worth arguing with anyone so obtuse as not to realise that this complex of discoveries is the greatest achievement of science in the twentieth century.' My misgiving about this engagingly calculated piece of arrogance is that I'd have a hard time defending it against a rival claim for, say, quantum theory or relativity.

Watson and Crick's was a digital revolution and it has gone exponential since 1953. You can read a gene today, write it out precisely on a piece of paper, put it in a library, then at any time in the future reconstitute that exact gene and put it back into an animal or plant. When the human genome project is completed, probably around 2003,* it will be possible to write the entire human genome on a couple of standard CDs, with enough space over for a large textbook of explanation. Send the boxed set of two CDs out into deep space and the human race can go extinct, happy in the knowledge that there is now at least a faint chance for an alien civilization to reconstitute a living human being. In one respect (though not in another), my speculation is at least more plausible than the plot of *Jurassic Park*. And both speculations rest upon the digital accuracy of DNA.

Of course, digital theory has been most fully worked out not by neurobiologists or geneticists, but by electronics engineers. The digital telephones, televisions, music reproducers and microwave beams of the late twentieth century are incomparably faster and more accurate than their analogue forerunners, and this is critically because they are digital. Digital computers are the

---

* It was indeed 2003 when it was formally declared complete, although there was still some tidying up to be done.

crowning achievement of this electronic age, and they are heavily implicated in telephone switching, satellite communications and data transmission of all kinds, including that phenomenon of the present decade, the World Wide Web. The late Christopher Evans summed up the speed of the twentieth-century digital revolution with a striking analogy to the car industry.

> Today's car differs from those of the immediate post-war years on a number of counts ... But suppose for a moment that the automobile industry had developed at the same rate as computers and over the same period: how much cheaper and more efficient would the current models be? If you have not already heard the analogy the answer is shattering. Today you would be able to buy a Rolls-Royce for £1.35, it would do three million miles to the gallon, and it would deliver enough power to drive the *Queen Elizabeth II*. And if you were interested in miniaturization, you could place half a dozen of them on a pinhead.

It is computers that make us notice that the twentieth century is the digital century; lead us to spot the digital in genetics, neurobiology and – though here I lack the confidence of knowledge – physics.

For it could be argued that quantum theory – the part of physics most distinctive of the twentieth century – is fundamentally digital. The Scottish chemist Graham Cairns-Smith tells how he was first exposed to this apparent graininess:

> I suppose I was about eight when my father told me that nobody knew what electricity was. I went to school the next day, I remember, and made this information generally available to my friends. It did not create the kind of sensation I had been banking on, although it caught the attention of one whose father worked at the local power station. His father actually made electricity so obviously he would know what it was. My friend promised to ask and report back. Well, eventually he did and I cannot say I was much impressed with the result. 'Wee sandy stuff' he said, rubbing his thumb and forefinger together to emphasise just how tiny the grains were. He seemed unable to elaborate further.

The experimental predictions of quantum theory are upheld to the tenth place of decimals. Any theory with such a spectacular grasp on reality commands our respect. But whether we conclude that the universe itself is grainy – or that discontinuity is forced upon an underlying deep continuity only when we try to measure it – I do not know; and physicists will sense that the matter is too deep for me.

It should not be necessary to add that this gives me no satisfaction. But sadly there are literary and journalistic circles in which ignorance or incomprehension of science is boasted with pride and even glee. I have made the point often enough to sound plaintive. So let me quote, instead, one of the most justly respected commentators on today's culture, Melvyn Bragg:

> There are still those who are affected enough to say they know nothing about the sciences as if this somehow makes them superior. What it makes them is rather silly, and it puts them at the fag end of that tired old British tradition of intellectual snobbery which considers all knowledge, especially science, as 'trade'.

Sir Peter Medawar, that swashbuckling Nobel Prize-winner whom I've already quoted, said something similar about 'trade':

> It is said that in ancient China the mandarins allowed their fingernails – or anyhow one of them – to grow so extremely long as manifestly to unfit them for any manual activity, thus making it perfectly clear to all that they were creatures too refined and elevated ever to engage in such employments. It is a gesture that cannot but appeal to the English, who surpass all other nations in snobbishness; our fastidious distaste for the applied sciences and for trade has played a large part in bringing England to the position in the world which she occupies today.

So, if I have difficulties with quantum theory, it is not for want of trying and certainly not a source of pride. As an evolutionist, I endorse Steven Pinker's view, that Darwinian natural selection has designed our brains to understand the slow dynamics of large objects on the African savannahs. Perhaps somebody should devise a computer game in which bats and balls behave according to a screened illusion of quantum dynamics. Children brought up on such a game might find modern physics no more impenetrable than we find the concept of stalking a wildebeest.

Personal uncertainty about the uncertainty principle reminds me of another hallmark that will be alleged for twentieth-century science. This is the century, it will be claimed, in which the deterministic confidence of the previous one was shattered. Partly by quantum theory. Partly by chaos (in the trendy, not the ordinary language, meaning). And partly by relativism (cultural relativism, not the sensible, Einsteinian meaning).

Quantum uncertainty and chaos theory have had deplorable effects upon popular culture, much to the annoyance of genuine aficionados. Both are regularly exploited by obscurantists, ranging from professional quacks to daffy New Agers. In America, the self-help 'healing' industry coins millions, and it has not been slow to cash in on quantum theory's formidable talent to bewilder. This has been documented by the American physicist Victor Stenger. One well-heeled healer wrote a string of best-selling books on what he calls 'Quantum Healing'. Another book in my possession has sections on quantum psychology, quantum responsibility, quantum morality, quantum aesthetics, quantum immortality and quantum theology.

Chaos theory, a more recent invention, is equally fertile ground for those with a bent for abusing sense. It is unfortunately named, for 'chaos' implies randomness. Chaos in the technical sense is not random at all. It is completely determined, but it depends hugely, in strangely hard-to-predict ways, on tiny differences in initial conditions. Undoubtedly it is mathematically interesting. If it impinges on the real world, it would rule out ultimate prediction. If the weather is technically chaotic, weather forecasting in detail becomes impossible. Major events like hurricanes might be determined by tiny causes in the past – such as the now proverbial flap of a butterfly's wing. This does not mean that you can flap the equivalent of a wing and hope to generate a hurricane. As the physicist Robert Park says, this is 'a total misunderstanding of what chaos is about . . . while the flapping of a butterfly's wings might conceivably trigger a hurricane, killing butterflies is unlikely to reduce the incidence of hurricanes'.

Quantum theory and chaos theory, each in its own peculiar ways, may call into question the predictability of the universe, in deep principle. This could be seen as a retreat from nineteenth-century confidence. But nobody really thought such fine details would ever be predicted in practice, anyway. The most confident determinist would always have admitted that, in practice, the sheer complexity of interacting causes would defeat accurate prediction of weather or turbulence. So chaos doesn't make a lot of difference in practice. Conversely, quantum events are statistically smothered, and massively so, in most realms that impinge on us. So the possibility of prediction is, for practical purposes, restored.

In the late twentieth century, prediction of future events in practice has never been more confident or more accurate. This is dramatic in the feats of space engineers. Previous centuries could predict the return of Halley's Comet. Twentieth-century science can hurl a projectile along the right trajectory to intercept it, precisely computing and exploiting the gravitational slings of the solar system.* Quantum theory itself, whatever the indeterminacy at its heart, is spectacularly accurate in the experimental accuracy of its predictions. The late Richard Feynman assessed this accuracy as equivalent to knowing the distance between New York and Los

* And less than a decade later, twenty-first-century science did just that, albeit for a different comet. In 2004 the European Space Agency launched the spacecraft Rosetta. Ten years and four billion miles later, after using the gravitation slingshot effect of Mars and then Earth (twice), and after close encounters with two large asteroids, Rosetta finally achieved orbit around its target, the comet 67P/Churyumov–Gerasimenko. Rosetta then launched the probe Philae, which successfully landed on the comet, using grappling harpoons to stop it bouncing off, the gravitational field of the comet being very weak.

Angeles to the width of one human hair. Here is no licence for anything-goes, intellectual flappers, with their quantum theology and quantum you-name-it.

Cultural relativism is the most pernicious of these myths of twentieth-century retreat from Victorian certainty. A modish fad sees science as only one of many cultural myths, no more true or valid than the myths of any other culture. Many in the academic community have discovered a new form of anti-scientific rhetoric, sometimes called the 'postmodern critique' of science. The most thorough whistle-blowing on this kind of thing is Paul Gross and Norman Levitt's splendid book *Higher Superstition: the academic left and its quarrels with science*. The American anthropologist Matt Cartmill sums up the basic credo:

> Anybody who claims to have objective knowledge about anything is trying to control and dominate the rest of us ... There are no objective facts. All supposed 'facts' are contaminated with theories, and all theories are infested with moral and political doctrines ... Therefore, when some guy in a lab coat tells you that such and such is an objective fact ... he must have a political agenda up his starched white sleeve.

There are even a few, but very vocal, fifth columnists within science itself who hold exactly these views, and use them to waste the time of the rest of us.

Cartmill's thesis is that there is an unexpected and pernicious alliance between the know-nothing fundamentalist religious right and the sophisticated academic left. A bizarre manifestation of the alliance is joint opposition to the theory of evolution. The opposition of the fundamentalists is obvious. That of the left is a compound of hostility to science in general, of 'respect' for tribal creation myths, and various political agendas. Both these strange bedfellows share a concern for 'human dignity' and take offence at treating humans as 'animals'. Moreover, in Cartmill's words,

> Both camps believe that the big truths about the world are moral truths. They view the universe in terms of good and evil, not truth

and falsehood. The first question they ask about any supposed fact is whether it serves the cause of righteousness.

And there is a feminist angle, which saddens me, for I am sympathetic to true feminism.

> Instead of exhorting young women to prepare for a variety of technical subjects by studying science, logic, and mathematics, Women's Studies students are now being taught that logic is a tool of domination ... the standard norms and methods of scientific inquiry are sexist because they are incompatible with 'women's ways of knowing'.* The authors of the prize-winning book with this title report that the majority of the women they interviewed fell into the category of 'subjective knowers', characterized by a 'passionate rejection of science and scientists'. These 'subjectivist' women see the methods of logic, analysis and abstraction as 'alien territory belonging to men' and 'value intuition as a safer and more fruitful approach to truth'.

That was a quotation from the historian and philosopher of science Noretta Koertge, who is understandably worried about a subversion of feminism which could have a malign influence upon women's education. Indeed, there is an ugly, hectoring streak in this kind of thinking. Barbara Ehrenreich and Janet McIntosh witnessed a woman psychologist speaking at an interdisciplinary conference. Various members of the audience attacked her use of the 'oppressive, sexist, imperialist, and capitalist scientific method. The psychologist tried to defend science by pointing to its great discoveries – for example, DNA. The retort came back: "You believe in DNA?"'

Fortunately, there are still many intelligent young women prepared to enter a scientific career, and I should like to pay tribute to their courage in the face of such bullying intimidation.[†]

---

* I commented on this sort of patronizing claptrap in the first piece in this collection; see footnote on page 23 above.

† Nor is it just women who are subject to this kind of bullying. In the footnote on page 88 I described the successful interception of

I have come so far with scarcely a mention of Charles Darwin. His life spanned most of the nineteenth century, and he died with every right to be satisfied that he had cured humanity of its greatest and grandest illusion. Darwin brought life itself within the pale of the explicable. No longer a baffling mystery demanding supernatural explanation, life, with the complexity and elegance that define it, grows and gradually emerges, by easily understood rules, from simple beginnings. Darwin's legacy to the twentieth century was to demystify the greatest mystery of all.

Would Darwin be pleased with our stewardship of that legacy, and with what we are now in a position to pass to the twenty-first century? I think he would feel an odd mixture of exhilaration and exasperation. Exhilaration at the detailed knowledge, the comprehensiveness of understanding, that science can now offer, and the polish with which his own theory is being brought to fulfilment. Exasperation at the ignorant suspicion of science, and the air-headed superstition, that still persist.

Exasperation is too weak a word. Darwin might justifiably be saddened, given our huge advantages over himself and his contemporaries, at how little we seem to have done to deploy our superior knowledge in our culture. Late twentieth-century civilization, Darwin would be dismayed to note, though imbued with and surrounded by the products and advantages of science,

a comet by the European Space Agency in 2014. One of the heroes of this heartstopping feat of human ingenuity was Dr Matt Taylor, an Englishman (this being the happier time when Britain was still a wholehearted partner in European enterprises). While announcing the achievement to the press, Dr Taylor wore a colourful shirt, a present from his girlfriend, and his doing so was deemed sexist. This trumped-up scandal of 'offence to women' eclipsed the news of one of the greatest engineering achievements of all time, and reduced Matt Taylor to tears and abject apology. I could hardly have imagined a more poignant illustration for the jeremiad portions of this lecture.

has yet to draw science into its sensibility. Is there even a sense in which we have slipped backwards since Darwin's co-discoverer, Alfred Russel Wallace, wrote *The Wonderful Century*, a glowing scientific retrospective on his era?

Perhaps there was undue complacency in late nineteenth-century science, about how much had been achieved and how little more advancement could be expected. William Thomson, the first Lord Kelvin, President of the Royal Society, pioneered the transatlantic cable – symbol of Victorian progress – and also the second law of thermodynamics – C. P. Snow's litmus of scientific literacy. Kelvin is credited with the following three confident predictions: 'Radio has no future.' 'Heavier than air flying machines are impossible.' 'X-rays will prove to be a hoax.'

Kelvin also gave Darwin a lot of grief by 'proving', using all the prestige of the senior science of physics, that the sun was too young to have allowed time for evolution. Kelvin, in effect, said: 'Physics argues against evolution, so your biology must be wrong.' Darwin could have retorted: 'Biology shows that evolution is a fact, so your physics must be wrong.' Instead, he bowed to the prevailing assumption that physics automatically trumps biology, and fretted. Twentieth-century physics, of course, showed Kelvin wrong by powers of ten. But Darwin did not live to see his vindication,* and he never had the confidence to tell the senior physicist of his day where to get off.

In my attacks on millenarian superstition, I must beware of Kelvinian overconfidence. Undoubtedly there is much that we still don't know. Part of our legacy to the twenty-first century must be unanswered questions, and some of them are big ones. The science of any age must prepare to be superseded. It would be arrogant and rash to claim our present knowledge as all there is to know. Today's commonplaces, such as mobile telephones, would have seemed to

---

* Pleasingly at the hands of, among others, his son, the mathematician and geophysicist Sir George Darwin. Three of Charles Darwin's sons were knighted, though their father never was.

previous ages pure magic. And that should be our warning. Arthur C. Clarke, distinguished novelist and evangelist for the limitless power of science, has said: 'Any sufficiently advanced technology is indistinguishable from magic.' This is Clarke's Third Law.

Maybe, some day in the future, physicists will fully understand gravity, and build an anti-gravity machine. Levitating people may one day become as commonplace to our descendants as jet planes are to us. So, if someone claims to have witnessed a magic carpet zooming over the minarets, should we believe him, on the grounds that those of our ancestors who doubted the possibility of radio turned out to be wrong? No, of course not. But why not?

Clarke's Third Law doesn't work in reverse. Given that 'any sufficiently advanced technology is indistinguishable from magic', it does not follow that 'any magical claim that anybody may make at any time is indistinguishable from a technological advance that will come some time in the future'.

Yes, there have been occasions when authoritative sceptics have come away with egg on their pontificating faces. But a far greater number of magical claims have been made and never vindicated. A few things that would surprise us today will come true in the future. But lots and lots of things will not come true in the future. History suggests that the very surprising things that do come true are in a minority. The trick is to sort them out from the rubbish – from claims that will forever remain in the realm of fiction and magic.

It is right that, at the end of our century, we should show the humility that Kelvin, at the end of his, did not. But it is also right to acknowledge all that we have learned during the past hundred years. The digital century was the best I could come up with, as a single theme. But it covers only a fraction of what twentieth-century science will bequeath. We now know, as Darwin and Kelvin did not, how old the world is. About 4.6 billion years. We understand what Alfred Wegener was ridiculed for suggesting: that the shape of geography has not always been the same. South America not only looks as if it might jigsaw neatly under the bulge of Africa. It once did exactly that, until they split apart some 125 million years ago. Madagascar once touched Africa on one side and India on the other. That was before India set off across the widening ocean and crashed into Asia to raise the Himalayas. The map of the world's continents has a time dimension, and we who are privileged to live in the plate tectonic age know exactly how it has changed, when, and why.

We know roughly how old the universe is, and, indeed, that it has an age, which is the same as the age of time itself, and less than twenty billion years. Having begun as a singularity with huge mass and temperature and very small volume, the universe has been expanding ever since. The twenty-first century will probably settle the question whether the expansion is to go on for ever, or go into reverse. The matter in the cosmos is not homogeneous, but is gathered into some hundred billion galaxies, each averaging a hundred billion stars. We can read the composition of any star in some detail, by spreading its light in a glorified rainbow. Among the stars, our sun is generally unremarkable. It is unremarkable, too, in having planets in orbit, as we know from detecting tiny rhythmic shifts in the spectrums of other stars.* There is no direct evidence that any other planets house life. If they do, such inhabited islands may be so scattered as to make it unlikely that one will ever encounter another.

We know in some detail the principles governing the evolution of our own island of life. It is a fair bet that the most fundamental principle – Darwinian natural selection – underlies, in some form, other islands of life, if any there be. We know that our kind of life is built of cells, where a cell is either a bacterium or a colony of bacteria. The detailed mechanics of our kind of life depend upon the near-infinite variety of shapes assumed by a special class of

---

* Other methods of detecting planets are now available, including the faint dimming of a star when undergoing a planetary transit. The tally of 'exoplanets' increases steadily, and now numbers over three thousand.

molecules called proteins. We know that those all-important three-dimensional shapes are exactly specified by a one-dimensional code, the genetic code, carried by DNA molecules which are replicated through geological time. We understand why there are so many different species, although we don't know how many. We cannot predict in detail how evolution will go in the future, but we can predict the general patterns that are to be expected.

Among the unsolved problems we shall bequeath to our successors, physicists such as Steven Weinberg will point to their 'dreams of a final theory', otherwise known as the grand universal theory (GUT) or theory of everything (TOE). Theorists differ about whether it will ever be attained. Those who think it will would probably date this scientific epiphany somewhere in the twenty-first century. Physicists famously resort to religious language when discussing such deep matters. Some of them really mean it. The others are at risk of being taken literally, when really they intend no more than I do when I say 'God knows' to mean that I don't.

Biologists will reach their grail of writing down the human genome, early in the next century. They will then discover that it is not so final as some once hoped. The human embryo project – working out how the genes interact with their environments, including each other, to build a body – may take at least as long to complete. But it too will probably be finished during the twenty-first century, and artificial wombs built, if these should be thought desirable.

I am less confident about what is for me, as for most biologists, the outstanding scientific problem that remains: the question of how the human brain works, especially the nature of subjective consciousness. The last decade of this century has seen a flurry of big guns take aim at it, including Francis Crick no less, and Daniel Dennett, Steven Pinker and Sir Roger Penrose. It is a big, profound problem, worthy of minds like these. Obviously I have no solution. If I had, I'd deserve a Nobel Prize. It isn't even clear what kind of a problem it is, and therefore what kind of a brilliant idea would constitute a solution. Some people think the problem of consciousness an illusion: there's nobody home, and no problem to be solved. But before Darwin solved the riddle of life's provenance, in the last century, I don't think anybody had clearly posed what sort of a problem it was. It was only after Darwin had solved it that most people realized what it had been in the first place. I do not know whether consciousness will prove to be a big problem, solved by a genius, or will fritter unsatisfactorily away into a series of small problems and non-problems.

I am by no means confident that the twenty-first century will solve the human mind. But if it does, there may be an additional by-product. Our successors may then be in a position to understand the paradox of twentieth-century science. On the one hand, our century arguably added as much new knowledge to the human store as all previous centuries put together; while on the other hand the twentieth century ended with approximately the same level of supernatural credulity as the nineteenth, and rather more outright hostility to science. With hope, if not with confidence, I look forward to the twenty-first century and what it may teach us.

Dolittle and Darwin*

I WISH I COULD say that my early childhood in East Africa turned me on to natural history in general and human evolution in particular. But it wasn't like that. I came to science late. Through books.

My childhood was as near an idyll as you could expect, given that I was sent away to boarding school at seven. I survived that experience as well as the next boy, which means pretty well (some tragic exceptions were lost in the bullied tail of the distribution), and my excellent schooling finally got me into Oxford, 'that Athens of my riper age'.† Home life was genuinely idyllic, first in Kenya, then Nyasaland (now Malawi), then England, on the family farm in Oxfordshire. We were not rich, but we weren't poor either. We had no television, but that was only because my parents thought, with some justice, that there were better ways to spend time. And we had books.

* In 2004 the literary agent and impresario of science John Brockman invited his unrivalled circle of intellectual correspondents to contribute to *When We Were Kids*, a collection of essays on 'How a child becomes a scientist'. Since I planned, one day, to write a proper autobiography (it ended up being split into two, *An Appetite for Wonder* and *Brief Candle in the Dark*), my essay for the Brockman collection did something different. I chose to extol one particular children's author who, I believe, influenced me.

† Dryden, notwithstanding his Cambridge education.

Maybe obsessive reading imprints a love of words in a child, and perhaps it later assists the craft of writing. In particular, I wonder whether the formative influence that eventually led to my becoming a zoologist might have been a children's book: Hugh Lofting's *The Adventures of Doctor Dolittle*, which I read again and again, along with its numerous sequels. This series of books did not turn me on to science in any direct sense, but Dr Dolittle was a scientist, the world's greatest naturalist, and a thinker of restless curiosity. Long before either phrase was coined, he was a role model who raised my consciousness.

John Dolittle was an amiable country doctor who turned from human patients to animals. Polynesia, his parrot, taught him to speak the languages of animals, and this single skill provided the plots of nearly a dozen books. Where other books for children (including today's Harry Potter series) profligately invoke the supernatural as a panacea for all difficulties, Hugh Lofting rationed himself to a single alteration of reality, as in science fiction. Dr Dolittle could speak to animals: from this all else followed. When he was appointed to run the post office of the West African kingdom of Fantippo, he recruited migrant birds into the world's first airmail service; small birds carried a single letter each, storks large parcels. When his ship needed a turn of speed to overtake the wicked slave trader Davy Bones, thousands of gulls gave him a tow – and a child's imagination soared.* When he got within range of the slaving vessel, a swallow's keen eyesight aimed his cannon to superhuman accuracy. When a man was framed for murder, Dr Dolittle persuaded the judge to allow the accused's bulldog to take the stand as the only witness to his innocence, having established his credentials as interpreter by talking to the judge's dog, who divulged embarrassing secrets only the dog could have known.

---

* I remember shamelessly plagiarizing this image in a school essay when I was about nine. My English teacher praised my imagination and predicted that I'd grow up to be a famous writer. Little did he know I stole it from Hugh Lofting.

The feats that Dr Dolittle could achieve because of this one facility – talking to animals – were frequently mistaken for supernatural by the doctor's enemies. Cast into an African dungeon to be starved into submission, Dr Dolittle grew fatter and jollier. Thousands of mice carried in food, one crumb at a time, with water in walnut shells and even fragments of soap so he could wash and shave. His terrified captors naturally put it down to witchcraft, but we, the child readers, were privy to the simple and rational explanation. The same salutary lesson was rammed home again and again through these books. It might look like magic, and the bad guys thought it was magic, but there was a rational explanation.

Many children have power dreams in which a magic spell or a fairy godmother or God himself comes to their aid. My dreams were of talking to animals and mobilizing them against the injustices that humanity (as I thought, under the influence of my animal-loving mother and Dr Dolittle) inflicted on them. What Dr Dolittle produced in me was an awareness of what we would now call 'speciesism': the automatic assumption that humans deserve special treatment over and above all other animals simply because we are human. Doctrinaire anti-abortionists who blow up clinics and murder good doctors turn out on examination to be rank speciesists. An unborn baby is by any reasonable standards less deserving of moral sympathy than an adult cow. The pro-lifer screams 'Murder!' at the abortion doctor and goes home to a steak dinner. No child brought up on Dr Dolittle could miss the double standard. A child brought up on the Bible most certainly could.

Moral philosophy aside, Dr Dolittle taught me not evolution itself but a precursor to understanding it: the non-uniqueness of the human species in the continuity of animals. Darwin himself expended great effort to the very same end. Parts of *The Descent of Man* and *The Expression of the Emotions* are devoted to narrowing the gulf between us and our animal cousins. What Darwin did for his adult Victorian readers, Dr Dolittle did for at least one small boy in the 1940s and 1950s. When I later came to read *The Voyage of the Beagle*, I fancied a resemblance between Darwin and Dolittle.

Dolittle's top hat and frock coat, and the style of ship that he incompetently sailed and usually wrecked, showed him to be a rough contemporary of Darwin. But that was only the beginning. The love of nature, the gentle solicitude towards all creation, the prodigious knowledge of natural history, the scribbled descriptions in notebook after notebook of amazing discoveries in exotic foreign parts: surely Dr Dolittle and the 'Philos' of the *Beagle* might have met in South America or on the floating island of Popsipetel (shades of plate tectonics), and they would have been soul brothers. Dolittle's Pushmi-Pullyu, an antelope with a horned head at both ends, was scarcely more incredible than some of the fossils and other specimens discovered by the young Darwin.* When Dolittle needed to cross a chasm in Africa, swarms of monkeys gripped one another by arms and legs to constitute a living bridge. Darwin would instantly have recognized the scene: the army ants he observed in Brazil do exactly the same thing. Darwin later investigated the remarkable habit among ants of taking slaves, and he, like Dolittle, was ahead of his time in his passionate hatred of slavery among humans. It was the only thing that roused both these normally mild naturalists to hot anger, in Darwin's case leading to a falling-out with Captain FitzRoy.

One of the most poignant scenes in all children's literature is in *Doctor Dolittle's Post Office*: Zuzanna, a West African woman whose husband has been seized by Davy Bones, the wicked slaver, is discovered all alone in a tiny canoe in mid-ocean, exhausted and weeping, bowed over her paddle after having given up her pursuit of the slaving vessel. She at first refuses to speak to the kindly doctor, assuming that any white man must be as evil as Davy Bones. But he coaxes her confidence and then summons up the resourceful fury of the animal kingdom in a successful campaign to overpower the slaver and rescue her husband. What irony that Hugh Lofting's

---

* Although I can't have been the only child to wonder how the Pushmi-Pullyu disposed of the waste products from the food that went in through its two mouths.

books are now banned as racist by sanctimonious public librarians! There is something in the charge. His drawings of Africans are steatopygic caricatures. Prince Bumpo, heir to the kingdom of the Jolliginki and an avid reader of fairy tales, saw himself as a Prince Charming but was convinced that his black face would frighten any Sleeping Beauty wakened by his kiss. So he persuaded Dr Dolittle to mix a special preparation to turn his face white. Not good consciousness-raising, by today's lights, and hindsight can find no excuse. But Hugh Lofting's 1920s simply were racist by today's standards,* and of course Darwin was too, like all Victorians, notwithstanding his hatred of slavery. Instead of being smugly censorious, we should look to our own accepted mores. Which of our unnoticed isms will the hindsight of future generations condemn? The obvious candidate is speciesism, and here Hugh Lofting's positive influence far outweighs the peccadillo of racial insensitivity.

Dr Dolittle resembles Charles Darwin also in his iconoclasm. Both are scientists who continually question accepted wisdom and conventional knowledge, because of their own temperament and also because they've been briefed by their animal informants. The habit of questioning authority is one of the most valuable gifts that a book, or a teacher, can give a young would-be scientist. Don't just accept what everybody tells you – think for yourself. I believe my childhood reading prepared me to love Charles Darwin when my adult reading finally brought him into my life.

* Some early Agatha Christies are worse but, as far as I know, unbanned. As for *Bulldog Drummond*, the 1920s equivalent of James Bond, he once had occasion to disguise himself as an African. The words with which he finally and dramatically revealed his true identity to the villain were: 'Every beard is not false, but every nigger smells. That beard ain't false, dearie, and dis nigger don't smell. So I'm thinking there's somethin wrong somewhere.' Prince Bumpo's ambitions to be a white Prince Charming seem mild by comparison.

II

ALL ITS MERCILESS GLORY

IF THE FIRST SECTION of this book was about what science *is*, the second homes in on science as it is *done* – specifically, on the development and refinement of Darwin's great theory, now established as scientific fact, of evolution by natural selection in, as Richard has put it elsewhere, 'all its merciless glory'.* The successive pieces illustrate how the theory was launched in an unlikely double-act of scientific gentlemanliness; how it works, and how far its power and validity might extend; how it is taken forward, and how it is misunderstood. Throughout runs the constant drive to refine, clarify and extend the application of this most powerful of scientific ideas.

The first piece, a speech delivered in the Linnean Society to commemorate the 1858 reading there of Charles Darwin's and Alfred Russel Wallace's papers that broke the news of their world-shaking discoveries, gives specific and poignant form to the values of science – and scientists – enumerated and defended in section I. After relating the joint work of the two great Victorian scientists, it ends with a bold supposition that Darwinian natural selection is the only adequate explanation not only for how life *has* evolved but for how it *could* evolve. Predecessors honoured, successors challenged: these are the hallmarks of Dawkinsian scientific discourse.

Successors challenged, and self challenged too. 'Universal Darwinism', written nearly twenty years earlier, subjects that bold

* In the foreword to David P. Hughes, Jacques Brodeur and Frédéric Thomas, *Host Manipulation by Parasites*.

supposition to a rigorous interrogation by way of a systematic review of the six alternative theories of evolution identified by the great German-American evolutionary biologist Ernst Mayr. It then goes one step further by setting out the stall for a new discipline of 'evolutionary exobiology'. A restless ambition in the cause of a passionately held conviction is not that rare; critical rigour is not that rare. The capacity to apply the latter to one's own ventures in the former is undoubtedly rarer than either; the evident enthusiasm for the task apparent here, perhaps rarer still. Its reward? The unambiguous assertiveness of a defending counsel confident he has established his case:

> Darwinism . . . is the only force I know that can, in principle, guide evolution in the direction of adaptive complexity. It works on this planet. It doesn't suffer from any of the drawbacks that beset the other five classes of theory, and there is no reason to doubt its efficacy throughout the universe.

When Darwin himself was laying down his theory, of course, the gene had not yet been identified, let alone pinpointed as the subject of natural selection. 'An ecology of replicators', first published in a collection in honour of Mayr, takes up the discourse of evolution in the context of twentieth-century debates about the level at which natural selection takes place, making the case for the gene, as the only *replicator* in the system, with exemplary clarity. Significantly, the core of the article is the investigation of an apparent disagreement (with Mayr himself) to identify where it is real and where illusory; the purpose, as so often, that of making a key *distinction* that enhances and refines understanding, and of revealing a key *commonality* under a disparity in terminology or expression.

A recurrent motif in this section is an insistence on the inadmissibility of group selection, the notion that the Darwinian principle can operate at the level of a family, a tribe, a species. 'Twelve misunderstandings of kin selection' is a *tour de force* in this campaign, a kind of scholarly sheepdog trial in which a series of meanderings from the proper path are patiently and efficiently redirected to the pen. One might expect this text, written as it is for a specialist periodical, and with a great deal of ground to cover, to be dry, bland, impersonal. Not so. Sentences such as the following reveal the kindred spirit of Douglas Adams: 'So it is that today the sensitive ethologist with his ear to the ground detects a murmuring of sceptical growls, rising to an occasional crescendo of smug baying when one of the early triumphs of the theory encounters new problems.' How many other writers for the esoteric end of the scientific bookshelf would dare such a flight? Equally characteristic is the 'Apology' at the end, emphasizing that the preceding critical explication is motivated not by any drive to score points over others, but by a desire to increase the common understanding. The progress of science trumps the triumph of the individual, every time.

G.S.

'More Darwinian than Darwin': the Darwin–Wallace papers*

I T IS IN the nature of scientific truths that they are waiting to be discovered, by whoever has the ability to do so. If two different people independently discover something in science, it will be the same truth. Unlike works of art, scientific truths do not change their nature in response to the individual human beings who discover them. This is both a glory, and a limitation, of science. If

* In 1858, Charles Darwin was startled to receive a manuscript from what was then the Federated Malay States, written by a little-known naturalist and collector, Alfred Russel Wallace. Wallace's paper laid out, in every particular, the theory of evolution by natural selection, the theory that Darwin had first thought of twenty years earlier. For reasons that are disputed, Darwin had not published his theory, although he wrote it out very fully in 1844. Wallace's letter threw Darwin into a tailspin of anxiety. He first thought he should cede priority to Wallace. However, his friends the geologist Charles Lyell and the botanist Joseph Hooker, two elder statesmen of British science, convinced him of a compromise. Wallace's 1858 paper and two earlier papers by Darwin would be read out at the Linnean Society in London, thereby achieving joint credit. In 2001, the Linnean Society decided to mount a plaque, on the very spot, to commemorate the historic event. I was invited to perform the unveiling, and this is a slightly abridged version of the speech with which I did so. The occasion felt celebratory. It was a pleasure to meet several members of both the Darwin and Wallace families, and in some cases introduce them to each other for the first time.

Shakespeare had never lived, nobody else would have written *Macbeth*. If Darwin had never lived, somebody else would have discovered natural selection. In fact, somebody did – Alfred Russel Wallace. And that is why we are here today.

On 1 July 1858 was launched upon the world the theory of evolution by natural selection, certainly one of the most powerful and far-reaching ideas ever to occur to a human mind. It occurred not to one mind, but two. Here I want to note that both Darwin and Wallace distinguished themselves not just for the discovery which they independently made, but for the generosity and humanity with which they resolved their priority in doing so. Darwin and Wallace seem to me to symbolize not just exceptional brilliance in science but the spirit of amicable cooperation which science, at its best, fosters.

The philosopher Daniel Dennett has written: 'Let me lay my cards on the table. If I were to give an award for the single best idea anyone has ever had, I'd give it to Darwin, ahead of Newton and Einstein and everyone else.' I have said something similar, although I didn't dare make the comparison with Newton and Einstein explicit. The idea we were talking about is, of course, evolution by natural selection. Not only is it the all but universally accepted explanation for the complexity and elegance of life; it is also, I strongly suspect, the only idea that in principle *could* provide that explanation.

But Darwin was not the only person who thought of it. When Professor Dennett and I made our remarks, we were – certainly in my case and I suspect that Dennett would agree – using the name Darwin to stand for 'Darwin and Wallace'. This happens to Wallace quite often, I am afraid. He tends to get a poor deal at the hands of posterity, partly through his own generous nature. It was Wallace himself who coined the word 'Darwinism', and he regularly referred to it as Darwin's theory. The reason we know Darwin's name more than Wallace's is that Darwin went on, a year later, to publish *On the Origin of Species*. The *Origin* not only explained and advocated the Darwin/Wallace theory of natural selection as the mechanism of evolution. It also – and this had to be done at book length – set out the multifarious evidence for the *fact* of evolution itself.

The drama of how Wallace's letter arrived at Down House on 17 June 1858, casting Darwin into an agony of indecision and worry, is too well known for me to retell it. In my view the whole episode is one of the more creditable and agreeable in the history of scientific priority disputes – precisely because it wasn't a dispute, although it so very easily could have become one. It was resolved amicably, and with heart-warming generosity on both sides, especially on Wallace's. As Darwin later wrote in his *Autobiography*,

> Early in 1856 Lyell advised me to write out my views pretty fully, and I began at once to do so on a scale three or four times as extensive as that which was afterwards followed in my *Origin of Species*; yet it was only an abstract of the materials which I had collected, and I got through about half the work on this scale. But my plans were overthrown, for early in the summer of 1858 Mr Wallace, who was then in the Malay archipelago, sent me an essay On the Tendency of Varieties to depart indefinitely from the Original Type; and this essay contained exactly the same theory as mine. Mr Wallace expressed the wish that if I thought well of his essay, I should send it to Lyell for perusal.
>
> The circumstances under which I consented at the request of Lyell and Hooker to allow of an extract from my MS., together with a letter to Asa Gray, dated September 5, 1857, to be published at the same time with Wallace's Essay, are given in the *Journal of the Proceedings of the Linnean Society*, 1858, p. 45. I was at first very unwilling to consent, as I thought Mr Wallace might consider my doing so unjustifiable, for I did not then know how generous and noble was his disposition. The extract from my MS. and the letter to Asa Gray . . . had neither been intended for publication, and were badly written. Mr Wallace's essay, on the other hand, was admirably expressed and quite clear. Nevertheless our joint productions excited very little attention, and the only published notice of them which I can remember was by Professor Haughton of Dublin, whose verdict was that all that was new in them was

false, and what was true was old. This shows how necessary it is that any new view should be explained at considerable length in order to arouse public attention.

Darwin was over-modest about his own two papers. Both are models of the explainer's art. Wallace's paper is also very clearly argued. His ideas were, indeed, remarkably similar to Darwin's own, and there is no doubt that Wallace arrived at them independently. In my opinion the Wallace paper needs to be read in conjunction with his earlier paper, published in 1855, in the *Annals and Magazine of Natural History*. Darwin read this paper when it came out. Indeed, it led to Wallace joining his large circle of correspondents, and to his engaging Wallace's services as a collector. But, oddly, Darwin did not see in the 1855 paper any warning that Wallace was by then a convinced evolutionist of a very Darwinian stamp. I mean as opposed to the Lamarckian view of evolution which saw modern species as all on a ladder, changing into one another as they moved up the rungs. By contrast Wallace, in 1855, had a clear view of evolution as a branching tree, exactly like Darwin's famous diagram which became the only illustration in *The Origin of Species*. The 1855 paper, however, makes no mention of natural selection or the struggle for existence.

That was left to Wallace's 1858 paper, the one which hit Darwin like a lightning bolt. Here, Wallace even used the phrase 'Struggle for Existence'. Wallace devotes considerable attention to the exponential increase in numbers (another key Darwinian point). Wallace wrote:

> The greater or less fecundity of an animal is often considered to be one of the chief causes of its abundance or scarcity; but a consideration of the facts will show us that it really has little or nothing to do with the matter. Even the least prolific of animals would increase rapidly if unchecked, whereas it is evident that the animal population of the globe must be stationary, or perhaps . . . decreasing.

Wallace deduced from this that:

> The numbers that die annually must be immense; and as the individual existence of each animal depends upon itself, those that die must be the weakest.

Wallace's peroration could have been Darwin himself writing:

> The powerful retractile talons of the falcon- and the cat-tribes have not been produced or increased by the volition of those animals; but among the different varieties which occurred in the earlier and less highly organized forms of these groups, those always survived longest which had the greatest facilities for seizing their prey ... Even the peculiar colours of many animals, especially insects, so closely resembling the soil or the leaves or the trunks on which they habitually reside, are explained on the same principle; for though in the course of ages varieties of many tints may have occurred, yet those races having colours best adapted to concealment from their enemies would inevitably survive the longest. We have also here an acting cause to account for that balance so often observed in nature, – a deficiency in one set of organs always being compensated by an increased development of some others – powerful wings accompanying weak feet, or great velocity making up for the absence of defensive weapons; for it has been shown that all varieties in which an unbalanced deficiency occurred could not long continue their existence. The action of this principle is exactly like that of the centrifugal governor of the steam engine, which checks and corrects any irregularities almost before they become evident.

The image of the steam governor is a powerful one which, I can't help feeling, Darwin might have envied.

Historians of science have raised the suggestion that Wallace's version of natural selection was not quite so Darwinian as Darwin himself believed. Wallace used the word 'variety' or 'race' as the level of entity at which natural selection acts. And some have suggested that Wallace, unlike Darwin who clearly saw selection as choosing among *individuals*, was proposing what modern theorists rightly denigrate as 'group selection'. This would be true if, by 'varieties', Wallace meant geographically separated groups or races of individuals. At first I wondered about this myself. But I believe a careful reading of Wallace's paper rules it out. I think that by 'variety' Wallace meant what we would nowadays call 'genetic type', even what a modern writer might mean by a gene. I think that, to Wallace in this paper, variety meant not local race of eagles, for example, but 'that set of individual eagles whose talons were hereditarily sharper than usual'.

If I am right, it is a similar misunderstanding to the one suffered by Darwin, whose use of the word 'race' in the subtitle of *The Origin of Species* is sometimes misread in support of racialism. That subtitle, or alternative title rather, is *The preservation of favoured races in the struggle for life*. Once again, Darwin was using 'race' to mean 'that set of individuals who share a particular hereditary characteristic', such as sharp talons, *not* a geographically distinct race such as the Hoodie Crow. If he had meant that, Darwin too would have been guilty of the group selection fallacy. I believe that neither Darwin nor Wallace was. And, by the same token, I do not believe that Wallace's conception of natural selection was different from Darwin's.

As for the calumny that Darwin plagiarized Wallace, that is rubbish. The evidence is very clear that Darwin did think of natural selection before Wallace, although he did not initially publish. We have his abstract of 1842 and his longer essay of 1844, both of which establish his priority clearly, as did his letter to Asa Gray of 1857 which was read here on the day we are celebrating. Why he delayed so long before publishing is one of the great mysteries of the history of science. Some historians have suggested that he was afraid of the religious implications, others the political ones. Maybe he was just a perfectionist.

When Wallace's letter arrived, Darwin was more surprised than we moderns might think he had any right to be. He wrote to Lyell: 'I never saw a more striking coincidence; if Wallace had had my manuscript sketch, written out in 1842, he could not have made a better short abstract of it. Even his terms now stand as Heads of my Chapters.'

The coincidence extended to both Darwin and Wallace being inspired by Robert Malthus on population. Darwin, by his own account, was immediately inspired by Malthus' emphasis on overpopulation and competition. He wrote in his autobiography:

> In October, 1838, that is, fifteen months after I had begun my systematic inquiry, I happened to read for amusement Malthus on population, and being well prepared to appreciate the struggle for existence which everywhere goes on from long continuous observation of the habits of animals and plants, it at once struck me that under these circumstances favourable variations would tend to be preserved and unfavourable ones to be destroyed. The result of this would be the formation of new species. Here, then, I had at last got a theory by which to work.

Wallace's epiphany was more delayed after his reading of Malthus, but was in a way more dramatic when it came . . . to his overheated brain in the midst of a malarial fever, on the island of Ternate in the Moluccas archipelago:

> I was suffering from a sharp attack of intermittent fever, and every day during the cold and succeeding hot fits had to lie down for several hours, during which time I had nothing to do but to think over any subjects then particularly interesting me . . .
>
> One day something brought to my recollection Malthus's 'Principles of Population'. I thought of his clear exposition of 'the positive checks to increase' – disease, accidents, war, and famine – which keep down the population of savage races to so much lower an average than that of more civilized peoples. It then occurred to me . . .

And Wallace proceeds to his own admirably clear exposition of natural selection.

There are other candidates for priority, apart from Darwin and Wallace. I'm not talking about the idea of evolution itself, of course;

there are numerous precedents there, including Erasmus Darwin. But for natural selection there are two other Victorians who have been championed – with something like the same zeal as Baconians show when disputing the authorship of Shakespeare. The two are Patrick Matthew and Edward Blyth; and Darwin himself mentions an even earlier one, W. C. Wells. Matthew complained that Darwin had overlooked him, and Darwin subsequently did mention him in later editions of the *Origin*. The following is from the introduction to the fifth edition:

> In 1831 Mr Patrick Matthew published his work on 'Naval Timber and Arboriculture', in which he gives precisely the same view of the origin of species as that . . . propounded by Mr Wallace and myself in the 'Linnean Journal', and as that enlarged in the present volume. Unfortunately the view was given by Mr Matthew very briefly in scattered passages in an Appendix to a work on a different subject, so that it remained unnoticed until Mr Matthew himself drew attention to it in the 'Gardener's Chronicle' . . .

As in the case of Edward Blyth, championed by Loren Eiseley, I think it is by no means clear that Matthew really did understand the importance of natural selection. The evidence is compatible with the view that these alleged predecessors of Darwin and Wallace saw natural selection as a purely negative force, weeding out misfits rather than building up the whole evolution of life (this, indeed, is a misconception under which modern creationists can be found labouring). I can't help feeling that, if you really understood that you were sitting on one of the greatest ideas ever to occur to a human mind, you would *not* bury it in scattered passages in an appendix to a monograph on naval timber. Nor subsequently choose the *Gardener's Chronicle* as the organ in which to claim your priority. That Wallace understood the enormous significance of what he had discovered, there is no doubt.

Darwin and Wallace did not remain always in total agreement. In old age, Wallace dabbled in spiritualism (in spite of his venerable appearance, Darwin never reached extreme old age), and from earlier times Wallace doubted that natural selection could account for the special abilities of the human mind. But the more important conflict between them came over sexual selection, and it has ramifications to this day, as Helena Cronin has documented in her beautifully written book *The Ant and the Peacock*. Wallace once said of himself: 'I am more Darwinian than Darwin himself.' He saw natural selection as ruthlessly utilitarian and he couldn't stomach Darwin's sexual selection interpretation of bird of paradise tails and similar bright coloration. Darwin's own stomach was not invulnerable. He wrote: 'The sight of a feather in a peacock's tail, whenever I gaze at it, makes me sick.' Nevertheless, Darwin reconciled himself to sexual selection, and became positively enthusiastic for it. Aesthetic whim, by females choosing among males, was enough to account for the peacock's tail and similar extravagances. Wallace hated this. So did just about everybody at the time except Darwin, sometimes for frankly misogynistic reasons. To quote Helena Cronin:

> Several authorities went further, emphasising the notorious fickleness of females. According to Mivart, 'Such is the instability of a vicious feminine caprice, that no constancy of coloration could be produced by its selective action.' Geddes and Thomson were of the gloomily misogynistic opinion that permanence of female taste was 'scarcely verifiable in human experience'.

Not for misogynistic reasons, Wallace strongly felt that female whim was not a proper explanation for evolutionary change. And Cronin uses his name for an entire strand of thought which lasts to this day. 'Wallaceans' are biased towards utilitarian explanations of bright coloration, while 'Darwinians' accept female whim as an explanation. Modern Wallaceans accept that peacocks' tails and similar bright organs are advertisements to females. But they want the males to be advertising genuine quality. A male with bright-coloured tail feathers is showing that he is a high-quality male. The Darwinian view of sexual selection, by contrast, is that the bright tail is valued by females for no additional qualities over and above the bright coloration itself. They like it because they like it because they like it. Females who choose attractive males have attractive sons who appeal to females of the next generation. Wallaceans more austerely insist that coloration must mean something useful.

The late W. D. Hamilton, my colleague at Oxford University, was a prime example of a Wallacean in this sense. He believed that sexually selected ornaments were badges of good health, selected for their capacity to advertise the health of a male – bad health as well as good.

One way to express Hamilton's Wallacean idea is to say that selection favours females who become skilled veterinary diagnosticians. At the same time, selection favours males who make it easy for them by, in effect, growing the equivalent of conspicuous thermometers and blood-pressure meters. The long tail of a bird of paradise, for Hamilton, is an adaptation to make it easy for females to diagnose the male's health, good or bad. An example of a good general diagnostic is a susceptibility to diarrhoea. A long dirty tail is a giveaway of ill-health. A long clean tail is the opposite. The longer the tail, the more unmistakable the badge of health, whether good health or poor. Obviously this honesty benefits the particular male only when his health is good. But Hamilton and other neo-Wallaceans have ingenious arguments* to the effect that natural selection favours honest badges in general, even if, in particular cases, honesty has painful consequences. Neo-Wallaceans believe that natural selection favours long tails precisely because they are an effective badge of health; both good health and (more paradoxically, but mathematical models of the theory really do stand up) poor health.

---

* I'm referring especially here to Alan Grafen's clever rendering into mathematical terms of qualitative arguments such as those of Amotz Zahavi. My own attempt to explain these matters is in the second edition of *The Selfish Gene*, written in a spirit of penance for the unjustified ridicule with which I had treated Zahavi's ideas in the first edition.

Sexual selectionists of the Darwin school also have their modern champions. Taking their line through R. A. Fisher in the first half of the twentieth century, modern Darwinian sexual selectionists have developed mathematical models which show that, also paradoxically, sexual selection governed by arbitrary female whim can lead to a runaway process such that the tail – or other sexually selected character – moves dangerously far away from its utilitarian optimum. The key to this family of theories is what modern geneticists call 'linkage disequilibrium'. When females choose, say, long-tailed males by whim, offspring of both sexes inherit their mother's whim genes and also their father's tail genes. It doesn't matter how arbitrary is the whim, the joint selection on both sexes can lead (at least if you do the mathematical theory in a certain way) to runaway evolution of longer tails, and of female preference for longer tails. So tails can become ludicrously long.

Cronin's elegant historical analysis shows that the Darwin/ Wallace opposition, in the field of sexual selection, persisted long after the deaths of the original protagonists, right through the twentieth century to today. It is especially pleasing – and might have amused the two men – that both the Darwinian and the Wallacean strands of sexual selection theory, more particularly in their modern forms, have a strong element of paradox. Both are capable of predicting surprising, even zany, sexual advertisements – which, indeed, we see in nature. The peacock's fan is only the most famous example.

I said the idea that occurred to Darwin and Wallace independently was one of the greatest, if not the greatest, ever to occur to a human mind. I want to end by giving this thought a universal spin. The opening words of my first book were:

> Intelligent life on a planet comes of age when it first works out the reason for its own existence. If superior creatures from space ever visit earth, the first question they will ask, in order to assess the level of our civilization, is: 'Have they discovered evolution yet?' Living organisms had existed on earth, without ever knowing why,

for over three thousand million years before the truth finally dawned on one of them. His name was Charles Darwin.

It would have been fairer, though less dramatic, to have said 'two of them' and to have coupled the name of Wallace with Darwin. But let me, in any case, pursue the universal perspective.

I believe the Darwin/Wallace theory of evolution by natural selection is the explanation not just of life on this planet, but of life in general. If life is ever found elsewhere in the universe, I make the prediction that, however different it may be in detail, there will be one important principle which it shares with our own form of life. It will have evolved, under the guidance of a mechanism broadly equivalent to the Darwin/Wallace mechanism of natural selection.

I am never quite sure how strongly to put this point.* The weak version, of which I am completely confident, is that no workable theory other than natural selection has ever been proposed. The strong form would be that no other workable theory ever *could* be proposed. Today, I think I'll stick with the weak form. It still has startling implications.

Natural selection not only explains everything we know about life. It does so with power, elegance and economy. It is a theory which has evident *stature*, a stature which really measures up to the magnitude of the problem which it sets out to solve.

Darwin and Wallace may not have been the first to get an inkling of the idea. But they were the first to understand the full magnitude of the problem, and the corresponding magnitude of the solution which jointly, and independently, occurred to them. This is the measure of their stature as scientists. The mutual generosity with which they settled the question of priority is the measure of their stature as human beings.

‖   * See the following essay in this collection, 'Universal Darwinism'.

Universal Darwinism*

IT IS WIDELY believed on statistical grounds that life has arisen many times all around the universe. However varied in detail alien forms of life may be, there will probably be certain principles that are fundamental to all life, everywhere. I suggest that prominent among these will be the principles of Darwinism. Darwin's theory of evolution by natural selection is more than a local theory to account for the existence and form of life on Earth. It is probably the only theory that *can* adequately account for the phenomena that we associate with life.

My concern is not with the details of other planets. I shall not speculate about alien biochemistries based on silicon chains, or alien neurophysiologies based on silicon chips. The universal perspective is my way of dramatizing the importance of Darwinism for our own biology here on Earth, and my examples will be mostly taken from Earthly biology. I do, however, also think that 'exobiologists' speculating about extraterrestrial life should make more use of evolutionary reasoning. Their writings have been rich in speculation about how extraterrestrial life might work, but poor in discussion about how it might *evolve*. This essay should, therefore,

* In 1982, one hundred years after Charles Darwin's death, his old university of Cambridge convened a centennial conference. What follows is a slightly edited version of the speech I gave at the conference, which appeared as a chapter in the book of proceedings, *Evolution from Molecules to Men*.

be seen firstly as an argument for the general importance of Darwin's theory of natural selection; secondly as a preliminary contribution to a new discipline of 'evolutionary exobiology'.

What Ernst Mayr called the 'growth of biological thought' is largely the story of Darwinism's triumph over alternative explanations of existence. The chief weapon of this triumph is usually portrayed as *evidence*. The thing that is said to be wrong with Lamarck's theory is that its assumptions are factually wrong. In Mayr's words: 'Accepting his premises, Lamarck's theory was as legitimate a theory of adaptation as that of Darwin. Unfortunately, these premises turned out to be invalid.' But I think we can say something stronger: *even accepting his premises*, Lamarck's theory is *not* as legitimate a theory of adaptation as that of Darwin because, unlike Darwin's, it is *in principle* incapable of doing the job we ask of it – explaining the evolution of organized, adaptive complexity. I believe this is so for all theories that have ever been suggested for the mechanism of evolution except Darwinian natural selection, in which case Darwinism rests on a securer pedestal than that provided by facts alone.

Now, I have made reference to theories of evolution 'doing the job we ask of them'. Everything turns on the question of what that job is. The answer may be different for different people. Some biologists, for instance, get excited about 'the species problem', while I have never mustered much enthusiasm for it as a 'mystery of mysteries'. For some, the main thing that any theory of evolution has to explain is the diversity of life – cladogenesis. Others may require of their theory an explanation of the observed changes in the molecular constitution of the genome.

I agree with John Maynard Smith that 'the main task of any theory of evolution is to explain adaptive complexity, i.e. to explain the same set of facts which the eighteenth-century theologian William Paley used as evidence of a Creator'. I suppose people like me might be labelled neo-Paleyists, or perhaps 'transformed Paleyists'. We concur with Paley that adaptive complexity demands a very special kind of explanation: either a Designer as Paley taught, or something such as natural selection that does the job of a designer.* Indeed, adaptive complexity is probably the best diagnostic of the presence of life itself.

Adaptive complexity as a diagnostic character of life

If you find something, anywhere in the universe, whose structure is complex and gives the strong appearance of having been designed

---

* I am astonished when I occasionally meet biologists who seem not to see the force of this. For example, the great Japanese geneticist Motoo Kimura was the main architect of the neutral theory of evolution. He was probably right that the majority of changes in gene frequency in populations (i.e. evolutionary changes) are not caused by natural selection but are neutral: new mutations come to dominate the population not because they are advantageous but because of random drift. The introduction to his great book *The Neutral Theory of Molecular Evolution* makes the concession that 'the theory does not deny the role of natural selection in determining the course of adaptive evolution'. But according to John Maynard Smith, Kimura was emotionally reluctant to make even this modest concession – so reluctant, indeed, that he couldn't bear to write it himself and asked his American colleague James Crow to write that one sentence for him! Kimura, and some other enthusiasts for the neutral theory, seem not to appreciate the importance of the functional near-perfection of biological adaptation. It's as though they've never seen a stick insect, a flying albatross, or a spider web. For them, the illusion of design is a trivial and rather dubious add-on, whereas for me and for those naturalists from whom I have learned (including Darwin himself) the complex perfection of biological design is the very core and centre of the life sciences. For us, the evolutionary changes that interested Kimura amount to resetting text in a different font. For us, what matters is not whether the text is written in Times New Roman or Helvetica. What matters is what the words mean. Kimura is probably right that only a minority of evolutionary changes are adaptive. But, for pity's sake, it is the minority that *matters*!

for a purpose, then that something either is alive, or was once alive, or is an artefact created by something alive. It is fair to include fossils and artefacts since their discovery on any planet would certainly be taken as evidence for life there.

Complexity is a statistical concept. A complex thing is a statistically improbable thing, something with a very low *a priori* likelihood of coming into being. The number of possible ways of arranging the 10^{27} atoms of a human body is obviously inconceivably large. Of these possible ways, only very few would be recognized as a human body. But this is not, by itself, the point. Any existing configuration of atoms is, *a posteriori*, unique; as 'improbable', with hindsight, as any other. The point is that, of all possible ways of arranging those 10^{27} atoms, only a tiny minority would constitute anything remotely resembling a machine that worked to keep itself in being, and to reproduce its kind. Living things are not just statistically improbable in the trivial sense of hindsight; their statistical improbability is limited by the *a priori* constraints of design. They are *adaptively* complex.

The term 'adaptationist' has been coined as a pejorative name for one who assumes, in Richard Lewontin's words, 'without further proof that all aspects of the morphology, physiology and behavior of organisms are adaptive optimal solutions to problems'. I have responded to this characterization elsewhere; here, I shall be an adaptationist in the much weaker sense that I shall only be *concerned* with those aspects of the morphology, physiology and behaviour of organisms that are undisputedly adaptive solutions to problems. In the same way, a zoologist may specialize on vertebrates without denying the existence of invertebrates. I shall be preoccupied with undisputed adaptations because I have defined them as my working diagnostic characteristic of all life, anywhere in the universe, in the same way as the vertebrate zoologist might be preoccupied with backbones because backbones are the diagnostic character of all vertebrates. From time to time I shall need an example of an undisputed adaptation, and the time-honoured eye will serve the purpose as well as ever – as it did, indeed, for Darwin himself, and for Paley: 'As far as the examination of the instrument goes, there is precisely the same proof that the eye was made for vision, as there is that the telescope was made for assisting it. They are made upon the same principles; both being adjusted to the laws by which the transmission and refraction of rays of light are regulated.'

If a similar instrument were found upon another planet, some special explanation would be called for. Either there is a God, or, if we are going to explain the universe in terms of blind physical forces, those blind physical forces are going to have to be deployed in a very peculiar way. The same is not true of non-living objects, as William Paley himself conceded.

A transparent pebble, polished by the sea, might act as a lens, focusing a real image. The fact that it is an efficient optical device is not particularly interesting because, unlike an eye or a telescope, it is too simple. We do not feel the need to invoke anything remotely resembling the concept of design. The eye and the telescope have many parts, all coadapted and working together to achieve the same functional end. The polished pebble has far fewer coadapted features: the coincidence of transparency, high refractive index and mechanical forces that polish the surface in a curved shape. The odds against such a threefold coincidence are not particularly great. No special explanation is called for.

Compare how a statistician decides what P value* to accept as

---

* Scientific experiments, especially in the biological sciences, are continually battling the suspicion that the result obtained might just be luck. Say a hundred patients are given an experimental medicine and compared with another hundred patients given a 'control', dummy pills that look the same but lack the active ingredient. If ninety of the experimental patients improve but only twenty of the controls, how do you know whether it is the drug that did it? Could it just be luck? Statistical tests exist to compute the probability that, if the drug really were doing nothing, you could have got the result you did get (or an even 'better' result) by luck. The 'P value' is that probability, and the lower it is, the less

evidence for an effect in an experiment. It is a matter of judgement and dispute, almost of taste, exactly when a coincidence becomes too great to stomach. But, no matter whether you are a cautious statistician or a daring statistician, there are some complex adaptations whose '*P* value', whose coincidence rating, is so impressive that nobody would hesitate to diagnose life (or an artefact designed by a living thing). My definition of living complexity is, in effect, 'that complexity which is too great to have come about through coincidence'. For the purposes of this essay, the problem that any theory of evolution has to solve is how living adaptive complexity comes about.

In his 1982 book *The Growth of Biological Thought*, Ernst Mayr helpfully lists what he sees as the six clearly distinct theories of evolution that have ever been proposed in the history of biology. I shall use this list to provide me with my main headings in this essay. For each of the six, instead of asking what the evidence is, for or against, I shall ask whether the theory is *in principle* capable of doing the job of explaining the existence of adaptive complexity. I shall take the six theories in order, and will conclude that only Theory 6, Darwinian selection, matches up to the task.

Theory 1. Built-in capacity for, or drive towards, increasing perfection

To the modern mind this is not really a theory at all, and I shall not bother to discuss it. It is obviously mystical, and does not explain anything that it does not assume to start with.

---

likely the result is to have been a matter of luck. Results with *P* values of 1 per cent or less are customarily taken as evidence, but that cut-off point is arbitrary. *P* values of 5 per cent may be taken as suggestive. For results that seem very surprising, for example an apparent demonstration of telepathic communication, a *P* value much lower than 1 per cent would be demanded.

Theory 2. Use and disuse plus inheritance of acquired characters

This is the Lamarckian theory. It is convenient to discuss it in two parts.

Use and disuse

It is an observed fact that on this planet living bodies sometimes become better adapted as a result of use. Muscles that are exercised tend to grow bigger. Necks that reach eagerly towards the treetops may lengthen in all their parts. Conceivably, if on some planet such acquired improvements could be incorporated into the hereditary information, adaptive evolution could result. This is the theory often associated with Lamarck, although there was more to what Lamarck said. Francis Crick, in 1982, said: 'As far as I know, no one has given *general* theoretical reasons why such a mechanism must be less efficient than natural selection.' In this section and the next I shall give two general theoretical objections to Lamarckism of exactly the kind which, I like to think, Crick was looking for. First, the shortcomings of the principle of use and disuse.

The problem is the crudity and imprecision of the adaptation that the principle of use and disuse is capable of providing. Consider the evolutionary improvements that must have occurred during the evolution of an organ such as an eye, and ask which of them could conceivably have come about through use and disuse. Does 'use' increase the transparency of a lens? No, photons do not wash it clean as they pour through it. The lens and other optical parts must have reduced, over evolutionary time, their spherical and chromatic aberration; could this come about through increased use? Surely not. Exercise might have strengthened the muscles of the iris, but it could not have built up the fine feedback control system which controls those muscles. The mere bombardment of a retina with coloured light cannot call colour-sensitive cones into existence, nor connect up their outputs so as to provide colour vision.

Darwinian types of theory, of course, have no trouble in

explaining all these improvements. Any improvement in visual accuracy could significantly affect survival. Any tiny reduction in spherical aberration may save a fast-flying bird from fatally misjudging the position of an obstacle. Any minute improvement in an eye's resolution of acute coloured detail may crucially improve its detection of camouflaged prey. The genetic basis of any improvement, however slight, will come to predominate in the gene pool. The relationship between selection and adaptation is a direct and close-coupled one. The Lamarckian theory, on the other hand, relies on a much cruder coupling: the rule that the more an animal uses a certain bit of itself, the bigger that bit ought to be. The rule occasionally might have some validity but not generally, and, as a sculptor of adaptation it is a blunt hatchet in comparison to the fine chisels of natural selection. This point is universal. It does not depend on detailed facts about life on this particular planet. The same goes for my misgivings about the inheritance of acquired characters.

Inheritance of acquired characters

The first problem here is that acquired characters are not always improvements. There is no reason why they should be, and indeed the majority of them are injuries. This is not just a fact about life on Earth. It has a universal rationale. If you have a complex and reasonably well-adapted system, the number of things you can do to it that will make it perform less well is vastly greater than the number of things you can do to it that will improve it. Lamarckian evolution will move in adaptive directions only if some mechanism – presumably selection – exists for distinguishing those acquired characters that are improvements from those that are not. Only the improvements should be imprinted into the germ line.

Although he was not talking about Lamarckism, Konrad Lorenz emphasized a related point for the case of learned behaviour, which is perhaps the most important kind of acquired adaptation. An animal learns to be a better animal during its own lifetime. It learns to eat sweet foods, say, thereby increasing its survival chances.* But there is nothing inherently nutritious about a sweet taste. Something, presumably natural selection, has to have built into the nervous system the arbitrary rule: 'treat sweet taste as reward', and this works because saccharine does not occur in nature whereas sugar does.

The principle holds, too, for morphological characters. Feet that are subjected to wear and tear grow tougher and more thick-skinned. The thickening of the skin is an acquired adaptation, but it is not obvious why the change went in this direction. In man-made machines, parts that are subjected to wear get thinner not thicker, for obvious reasons. Why does the skin on the feet do the opposite? Because natural selection has worked in the past to ensure an adaptive rather than a maladaptive response to wear and tear.

The relevance of this for would-be Lamarckian evolution is that there has to be a deep Darwinian underpinning even if there is a Lamarckian surface structure: a Darwinian choice of which potentially acquirable characters shall in fact be acquired and inherited. Lamarckian mechanisms cannot be fundamentally responsible for adaptive evolution. Even if acquired characters are inherited on some planet, evolution there will still rely on a Darwinian guide for its adaptive direction.

Theory 3. Direct induction by the environment

Adaptation, as we have seen, is a fit between organism and environment. The set of conceivable organisms is wider than the actual set. And there is a set of conceivable environments wider than the actual set. These two subsets match each other to some

---

* In a wild state, that is, where refined sugar doesn't exist except in the rare and painfully won case of honey. As it happens, the sweet tooth was an unfortunate example to have chosen because, in our domesticated world, the taste for sugar does not improve our chances of survival.

extent, and the matching is adaptation. We can re-express the point by saying that information from the environment is present in the organism. In a few cases this is vividly literal – a frog carries a picture of its environment around on its back. Such information is usually carried by an animal in the less literal sense that a trained observer, dissecting a new animal, can reconstruct many details of its natural environment.*

Now, how could the information get from the environment into the animal? Lorenz argues that there are two ways, natural selection and reinforcement learning, but that these are both *selective* processes in the broad sense.† There is, in theory, an alternative method for the environment to imprint its information on the organism, and that is by direct 'instruction'. Some theories of how the immune system works are 'instructive': antibody molecules are thought to be shaped directly by moulding themselves around antigen molecules. The currently favoured theory is, by contrast, selective. I take 'instruction' to be synonymous with the 'direct induction by the environment' of Mayr's Theory 3. It is not always clearly distinct from Theory 2.

Instruction is the process whereby information flows directly from its environment into an animal. A case could be made for treating imitation learning, latent learning and imprinting as instructive, but for clarity it is safer to use a hypothetical example. Think of an animal on some planet, deriving camouflage from its tiger-like stripes. It lives in long dry 'grass', and its stripes closely match the typical thickness and spacing of local grass blades. On our own planet such adaptation would come about through the selection of random genetic variation, but on the imaginary planet

* I later tried to make this idea more vivid, using the phrase 'Genetic Book of the Dead', referred to in several other essays in this collection.

† The psychologist B. F. Skinner was keen to emphasize the same point.

it comes about through direct instruction. The animals go brown except where their skin is shaded from the 'sun' by blades of grass. Their stripes are therefore adapted with great precision, not just to any old habitat, but to the precise habitat in which they have sunbathed, and it is this same habitat in which they are going to have to survive. Local populations are automatically camouflaged against local grasses. Information about the habitat, in this case about the spacing patterns of the grass blades, has flowed into the animals, and is embodied in the spacing pattern of their skin pigment.

Instructive adaptation demands the inheritance of acquired characters if it is to give rise to permanent or progressive evolutionary change. 'Instruction' received in one generation must be 'remembered' in the genetic (or equivalent) information. This process is in principle cumulative and progressive. However, if the genetic store is not to become overloaded by the accumulations of generations, some mechanism must exist for discarding unwanted 'instructions', and retaining desirable ones. I suspect that this must lead us, once again, to the need for some kind of selective process.

Imagine, for instance, a form of mammal-like life in which a stout 'umbilical nerve' enabled a mother to 'dump' the entire contents of her memory in the brain of her fetus. The technology is available even to our nervous systems: the corpus callosum can shunt large quantities of information from right hemisphere to left. An umbilical nerve could make the experience and wisdom of each generation automatically available to the next, and this might seem very desirable. But without a selective filter, it would take few generations for the load of information to become unmanageably large. Once again we come up against the need for a selective underpinning. I will leave this now, and make one more point about instructive adaptation (which applies equally to all Lamarckian types of theory).

The point is that there is a logical link-up between the two major theories of adaptive evolution – selection and instruction – and the two major theories of embryonic development – epigenesis and preformationism.* Instructive evolution can work only if embryology is preformationistic. If embryology is epigenetic, as it is on our planet, instructive evolution cannot work.

Briefly, if acquired characters are to be inherited, embryonic processes must be reversible: phenotypic change has to be read back into the genes (or equivalent). If embryology is preformationistic – the genes are a true blueprint – then it may indeed be reversible. You can translate a house back into its blueprint. But if embryonic

---

* There wasn't time during my Cambridge lecture to define these two historic views of how embryology might work; and the Cambridge audience commemorating Darwin wouldn't have needed the definition anyway. Preformationists propose that each generation contains the form of the next, either literally (a miniature body curled up in sperm or egg) or in coded form, as some kind of blueprint. Epigenesis means that each generation contains instructions for making the next generation, not as a blueprint but as something akin to a recipe or computer program. One could imagine a planet where embryology was preformationistic, as follows. The body of the parent is scanned, slice by slice, to build up instructions delivered to something equivalent to a 3D printer which then 'prints' the child, a copy of the parent's body; this copy then, if necessary, 'inflates' to full size. That isn't how embryology works on our planet, but it is the kind of way embryology would have to work for the hypothetical tiger-striped alien. Our kind of embryology on this planet is epigenetic; DNA is not a blueprint, contrary to most biology textbooks. It is a series of instructions, like a computer program, or a recipe or a sequence of origami paper-folds which, when followed, yields a body. Blueprint embryologies, if they exist, would be reversible, just as one can reconstruct the original architect plans by making measurements on a house. In no sense whatsoever is the body of the parent copied to make the child. Rather, the genes that made the body of the parent are copied (half of them, together with half from the other parent), and handed over as instructions to make the body of the next generation, and as unadulterated instructions to pass on to the grandchild generation. Bodies don't beget bodies. DNA begets bodies, and DNA begets DNA.

development is epigenetic; if, as on this planet, the genetic information is more like a recipe for a cake than a blueprint for a house, it is irreversible. There is no one-to-one mapping between bits of genome and bits of phenotype, any more than there is mapping between crumbs of cake and words of recipe. The recipe is not a blueprint that can be reconstructed from the cake. The transformation of recipe into cake cannot be put into reverse, and nor can the process of making a body. Therefore acquired adaptations cannot be read back into the 'genes' on any planet where embryology is epigenetic.

This is not to say that there could not, on some planet, be a form of life whose embryology was preformationistic. That is a separate question. How likely is it? The form of life would have to be very different from ours, so much so that it is hard to visualize how it might work. As for reversible embryology itself, it is even harder to visualize. Some mechanism would have to scan the detailed form of the adult body, carefully noting down, for instance, the exact location of brown pigment in a sun-striped skin, perhaps turning it into a linear stream of code numbers, as in a television camera. Embryonic development would read the scan out again, like a television receiver. I have an intuitive hunch that there is an objection in principle to this kind of embryology, but I cannot at present formulate it clearly.* All I am saying here is that, if planets are divided into those where embryology is preformationistic and those, like Earth, where embryology is epigenetic, Darwinian evolution could be supported on both kinds of planet, but Lamarckian evolution, even if there were not other reasons for doubting its

---

* Nowadays I'd have a stab at formulating it. For a start, it would be vulnerable to the problem already mentioned about wear and tear. A 'scan' of the parent's body would faithfully reproduce every scar, broken limb and missing foreskin, along with 'good' acquisitions such as toughened soles and learned wisdom. Once again there would be a need for a selective choice of 'good' acquisitions versus scars and the like. And what could the 'selector' be but some version of what Darwin proposed?

existence, could be supported only on the preformationistic planets – if there are any.

Theory 4. Saltationism

The great virtue of the idea of evolution is that it explains, in terms of blind physical forces, the existence of undisputed adaptations whose functionally directed statistical improbability is enormous, without recourse to the supernatural or mystical. Since we *define* an undisputed adaptation as an adaptation that is too complex to have come about by chance, how is it possible for a theory to invoke only blind physical forces in explanation? The answer – Darwin's answer – is astonishingly simple when we consider how self-evident Paley's Divine Watchmaker must have seemed to his contemporaries. The key is that the coadapted parts do not have to be assembled *all at once*. They can be put together in small stages. But they really do have to be *small* stages. Otherwise we are back again with the problem we started with: the creation by chance of complexity that is too great to have been created by chance!

Take the eye again, as an example of an organ that contains a large number of independent coadapted parts, say N. The *a priori* probability of any one of these N features coming into existence by chance is low, but not incredibly low. It is comparable to the chance of a crystal pebble being washed by the sea so that it acts as a rounded lens. Any one adaptation on its own could, plausibly, have come into existence through blind physical forces. If each of the N coadapted features confers some slight advantage on its own, then the whole many-parted organ can be put together over a long period of time. This is particularly plausible for the eye – ironically in view of that organ's niche of honour in the creationist pantheon. The eye is, *par excellence*, a case where a fraction of an organ is better than no organ at all; an eye without a lens or even a pupil, for instance, could still detect the looming shadow of a predator.

To repeat, the key to the Darwinian explanation of adaptive complexity is the replacement of instantaneous, coincidental, multi-dimensional luck, by gradual, inch by inch, smeared-out luck. Luck is involved, to be sure. But a theory that bunches the luck up into major steps is more incredible than a theory that spreads the luck out in small stages. This leads to the following general principle of universal biology. Wherever in the universe adaptive complexity shall be found, it will have come into being gradually through a series of small alterations, never through large and sudden increments in adaptive complexity.* We must reject Mayr's fourth theory, saltationism, as a candidate for explanation of the evolution of complexity.

It is almost impossible to dispute this rejection. It is implicit in the definition of adaptive complexity that the only alternative to gradualistic evolution is supernatural magic. This is not to say that the argument in favour of gradualism is a worthless tautology, an unfalsifiable dogma of the sort that creationists and philosophers are so fond of jumping about on. It is not *logically* impossible for a full-fashioned eye to spring *de novo* from virgin bare skin. It is just that the possibility is statistically negligible.

Now, it has recently been widely and repeatedly publicized that some modern evolutionists reject 'gradualism', and espouse what John Turner has rather irreverently called theories of evolution by jerks. Since these are reasonable people without mystical leanings, they must be gradualists in the sense in which I am here using the term: the 'gradualism' that they oppose must be defined differently. There are actually two confusions of language here, and I intend to

---

* I later used the metaphor of *Climbing Mount Improbable* in the book of that name. A complex piece of well-designed machinery such as an eye sits high on a peak of Mount Improbable. One side of the mountain is a sheer precipice, impossible to scale in a single leap – saltation. But on the other side of the mountain is a gentle slope, easy to climb simply by placing one foot in front of the other.

clear them up in turn. The first is the common confusion between 'punctuated equilibrium' and true saltationism.* The second is a confusion between two theoretically distinct kinds of saltation.

* Punctuated equilibrium, which briefly became familiar enough to be affectionately abbreviated to 'punk eek', was a theory advanced by the distinguished paleontologists Niles Eldredge and Stephen Jay Gould to account for apparent jumpiness of the fossil record. Unfortunately, partly abetted by Gould's persuasive but misleading rhetoric, the phrase later came to confuse three entirely different kinds of jump: first, macro-mutations or saltations (mutations of large effect, producing in extreme cases 'monstrosities' or 'hopeful monsters'); second, mass extinctions (like the sudden demise of the dinosaurs which cleared the stage for the mammals); and third (the meaning Eldredge and Gould intended as their original contribution) *rapid gradualism*. Eldredge and Gould together with some other paleontologists suggested, plausibly enough, that evolution stands still ('stasis') for long periods punctuated by sudden rapid bursts called 'speciation events'. They here invoked the theory of 'allopatric speciation'.

Allopatric speciation means the splitting of a species into two by reason of an initial geographic separation – for example, on islands or on opposite sides of a river or mountain range. While separated, the two populations have the opportunity to evolve away from each other so that, when they meet again in the fullness of time, they can no longer interbreed and are therefore defined as separate species. When a sub-population is separated from the mainland population on an offshore island, evolutionary change under island conditions can be so rapid that a new species arises almost instantaneously by the leisurely standards of geological time. An 'island', as discussed in 'The giant tortoise's tale' (page 341), doesn't have to be land surrounded by water. For a fish, a lake is an island. For an Alpine marmot, a high peak is an island. For ease of illustration, I'll continue to assume land surrounded by water.

When members of the island species migrate back to the mainland where the parent species is unchanged they will seem, to a paleontologist digging through the mainland rocks, to have arisen from the parent species in a single jump. The jump is an

Punctuated equilibrium is not macro-mutation, not saltation at all in the traditional sense of the term. It is, however, necessary to discuss it here, because it is popularly regarded as a theory of saltation, and its partisans quote, with approval, Huxley's criticism of Darwin for upholding the principle of *Natura non facit saltum*.* The punctuationist theory is portrayed as radical and revolutionary and at variance with the 'gradualistic' assumptions of both Darwin and the neo-Darwinian synthesis. Punctuated

_____

illusion. Gradual evolution really happened, albeit fast, and albeit offshore where the paleontologist isn't digging. It's easy to see that such 'rapid gradualism' is leagues away from true saltation. Yet Gould's rhetoric contrived to mislead a generation of students and laypeople into confusion with true saltationism – even into confusion with mass extinction and the consequent 'sudden' rise of new evolutionary blossomings such as that of the mammals after the death of the dinosaurs. This is an example of what I call 'poetic science' – a phrase I'll return to in the Afterword to this essay.

* 'Nature doesn't jump.' In Huxley's time his readers (including Darwin, whom he was directly addressing in a letter when he used the phrase) would have been, however unwillingly (especially in Darwin's case), schooled in Latin.

Stephen Gould himself was present at my Cambridge lecture. Afer it, he jumped up and mentioned saltationism as one of several historical alternatives to Darwinian selection. Did he really not understand the impossibility of explaining the complex illusion of design by saltation – leaping from the bottom to the top of Mount Improbable in a single bound? It seems hard to credit. Gould was deeply interested in, and knowledgeable about, history. He was historically correct in asserting that some scientists in the early twentieth century had espoused saltationism as (what they thought was) an alternative to gradualism. But he was scientifically – even logically – incorrect to say that saltationism could ever be a viable alternative to gradualism as an explanation for complex adaptation. In other words, the historical figures that he correctly cited were scientifically wrong; it was always obvious. It should have been obvious, even in their own time, that they were wrong; and Gould should have said so.

equilibrium, however, was originally conceived as what the orthodox neo-Darwinian synthetic theory should truly predict, on a paleontological timescale, if we take its embedded ideas of allopatric speciation seriously. It derives its 'jerks' by taking the 'stately unfolding' of the neo-Darwinian synthesis, and *inserting* long periods of stasis separating brief bursts of gradual, albeit rapid, evolution.

The plausibility of such 'rapid gradualism' is dramatized by a thought experiment of Ledyard Stebbins.* He imagines a species of mouse, evolving larger body size at such an imperceptibly slow rate that the differences between the means of successive generations would be utterly swamped by sampling error. Yet even at this slow rate Stebbins' mouse lineage would attain the body size of a large elephant in about sixty thousand years, a timespan so short that it would be regarded as instantaneous by paleontologists. Evolutionary change too *slow* to be detected by micro-evolutionists can nevertheless be too *fast* to be detected by macro-evolutionists.[†] What a paleontologist sees as a 'saltation' can in fact be a smooth

---

* Stebbins was an American botanist, revered as one of the founding fathers of the neo-Darwinian synthesis of the 1930s and 1940s.

† Nowadays I would hesitate to use these two terms, because they have been co-opted by creationists eager, as ever, to misuse scientific terms in order to deceive. Geneticists studying populations in the field are looking at micro-evolution. Paleontologists studying fossils through the ages are looking at macro-evolution. Macro-evolution is actually nothing more than what you get when micro-evolution goes on for a very long time. Creationists, with some inadvertent help from a few biologists who should have been on their guard, elevate the distinction to qualitative heights. They accept micro-evolution, such as the replacement of light-coloured peppered moths by dark mutants in the population. But they think macro-evolution is qualitatively, radically different. For a fuller treatment of the actual, and supposed, distinctions, see 'The "Alabama Insert"' in section IV of this collection.

and gradual change so slow as to be undetectable to the micro-evolutionist. This kind of paleontological 'saltation' has nothing to do with the one-generation macro-mutations that, I suspect, Huxley and Darwin had in mind when they debated *Natura non facit saltum*. Confusion has arisen here, possibly because some individual champions of punctuated equilibrium have also, incidentally, championed macro-mutation. Other 'punctuationists' have either confused their theory with macro-mutationism, or have explicitly invoked macro-mutation as one of the mechanisms of punctuation.

Turning to macro-mutation, or true saltation, the second confusion I want to clear up is between two kinds of macro-mutation that we might conceive of. I could name them, unmemorably, saltation (1) and saltation (2), but instead I shall pursue an earlier fancy for airliners as metaphors, and label them 'Boeing 747' and 'Stretched DC-8' saltation. 747 saltation is the inconceivable kind. It gets its name from Sir Fred Hoyle's much-quoted metaphor for his own cosmic misunderstanding of Darwinism. Hoyle compared Darwinian selection to a tornado, blowing through a junkyard and assembling a Boeing 747 (what he overlooked, of course, was the point about luck being 'smeared out' in small steps – see above). Stretched DC-8 saltation is quite different. It is not in principle hard to believe in at all. It refers to large and sudden changes in *magnitude* of some biological measure, without an accompanying large increase in adaptive information. It is named after an airliner that was made by elongating the fuselage of an existing design, not adding significant new complexity.* The

---

* The complexity that was inserted – seats, bulkheads, call buttons, tray tables etc. – was simply duplicated from the pre-stretched version of the plane. The biological parallel would be an increase in the number of vertebrae, with associated ribs, nerves, blood vessels etc., when a mutant snake has more segments than its parents. Such 'Stretched DC-8' evolutionary changes must have happened often, because, for example, different species of snake vary greatly in how many segments they have. Children

change from DC-8 to Stretched DC-8 is a big change in magnitude – a saltation, not a gradualistic series of tiny changes. But, unlike the change from junk-heap to 747, it is not a big increase in information content or complexity, and that is the point I am emphasizing by the analogy.

An example of DC-8 saltation would be the following. Suppose the giraffe's neck shot out in one spectacular mutational step. Two parents had necks of normal antelope length. They had a freak child with a neck of modern giraffe length, and all giraffes are descended from this freak. This is unlikely to be true on Earth,* but something like it may happen elsewhere in the universe. There is no objection to it in principle, in the sense that there is a profound objection to the (747) idea that a complex organ like an eye could arise from bare skin by a single mutation. The crucial difference is one of complexity.

I am assuming that the change from short antelope's neck to long giraffe's neck is *not* an increase in complexity. To be sure, both necks are exceedingly complex structures. You couldn't go from *no* neck to either kind of neck in one step; that would be 747 saltation. But once the complex organization of the antelope's neck already exists, the step to giraffe's neck is just an elongation: various things have to grow faster at some stage in embryonic development; existing complexity is preserved. In practice, of course, such a drastic change in magnitude would be highly likely to have deleterious repercussions which would render the macro-mutant unlikely to survive. The existing antelope heart probably could not pump the blood up to the newly elevated giraffe head. Such practical

---

must have been born with different whole numbers of segments from their parents, because you can't have a snake with a fraction of a vertebra.

* Actually, we have an intermediate in the form of the beautiful okapi, a cousin of the giraffes with a neck of intermediate length. But set that on one side for the sake of the example.

objections to evolution by DC-8 saltation can only help my case in favour of gradualism, but I still want to make a separate, and more universal, case against 747 saltation.

It may be argued that the distinction between 747 and DC-8 saltation is impossible to draw in practice. After all, DC-8 saltations, such as the proposed macro-mutational elongation of the giraffe's neck, may appear very complex: muscle blocks, vertebrae, nerves, blood vessels, all have to elongate together. Why does this not make it a 747 saltation, and therefore rule it out?

We know that single mutations can orchestrate changes in growth rates of many diverse parts of organs, and, when we think about developmental processes, it is not in the least surprising that this should be so. When a single mutation causes a *Drosophila* to grow a leg where an antenna ought to be, the leg grows in all its formidable complexity. But this is not mysterious or surprising, not a 747 saltation, because the organization of a leg is already present in the body before the mutation. Wherever, as in embryogenesis, we have a hierarchically branching tree of causal relationships, a small alteration at a senior node of the tree can have large and complex ramified effects on the tips of the twigs. But although the change may be large in magnitude, there can be no large and sudden increments in adaptive information. If you think you have found a particular example of a large and sudden increment in adaptively complex information in practice, you can be certain the adaptive information was already there, even if it is an atavistic 'throwback' to an earlier ancestor.

There is not, then, any objection in principle to theories of evolution by jerks, even the theory of hopeful monsters, provided that it is DC-8 saltation, not 747 saltation that is meant. No educated biologist actually believes in 747 saltation, but not all have been sufficiently explicit about the distinction between DC-8 and 747 saltation. An unfortunate consequence is that creationists and their journalistic fellow-travellers have been able to exploit saltationist-sounding statements of respected biologists. The biologist's intended meaning may have been what I am calling DC-8 saltation, or even non-saltatory punctuation; but the creationist *assumes* saltation in the sense that I have dubbed 747, and 747 saltation would, indeed, be a blessed miracle.

I also wonder whether an injustice is being done to Darwin, owing to this same failure of critics to come to grips with the distinction between DC-8 and 747 saltation. It is frequently alleged that Darwin was wedded to gradualism, and therefore that, if some form of evolution by jerks is proved, Darwin will have been shown wrong. This is undoubtedly the reason for the ballyhoo and publicity that has attended the theory of punctuated equilibrium. But was Darwin really opposed to all jerks? Or was he, as I suspect, strongly opposed only to 747 saltation?

As we have already seen, punctuated equilibrium has nothing to do with saltation, but anyway I think it is not at all clear that, as is often alleged, Darwin would have been discomfited by punctuationist interpretations of the fossil record. The following passage, from later editions of the *Origin*, sounds like something from a current issue of *Paleobiology*: 'The periods during which species have been undergoing modification, though very long as measured by years, have probably been short in comparison with the periods during which these same species remained without undergoing any change.' I believe we can reach a better understanding of Darwin's general gradualistic bias if we invoke the distinction between 747 and DC-8 saltation.

Perhaps part of the problem is that Darwin himself did not have the distinction. In some anti-saltation passages it seems to be DC-8 saltation that he has in mind. But on those occasions he does not seem to feel very strongly about it: 'About sudden jumps,' he wrote in a letter in 1860, 'I have no objection to them – they would aid me in some cases. All I can say is, that I went into the subject and found no evidence to make me believe in jumps [as a source of new species] and a good deal pointing in the other direction.' This does not sound like a man fervently opposed, in principle, to sudden jumps. And of course there is no reason why he *should* have been fervently opposed, if he only had DC-8 saltations in mind.

But at other times he really is pretty fervent, and on those occasions, I suggest, he is thinking of 747 saltation. As the historian Neal Gillespie puts it: 'For Darwin, monstrous births, a doctrine favored by Chambers, Owen, Argyll, Mivart, and others, from clear theological as well as scientific motives, as an explanation of how new species, or even higher taxa, had developed, was no better than a miracle: "it leaves the case of the co-adaptation of organic beings to each other and to their physical conditions of life, untouched and unexplained". It was "no explanation" at all, of no more scientific value than creation "from the dust of the earth"'.

Darwin's hostility to monstrous saltation, then, makes sense if we assume that he was thinking in terms of 747 saltation – the sudden invention of new adaptive complexity. It is highly likely that that is what he was thinking of, because that is exactly what many of his opponents had in mind. Saltationists such as the Duke of Argyll (though presumably not Huxley) wanted to believe in 747 saltation, precisely because it *did* demand supernatural intervention. Darwin did not believe in it, for exactly the same reason.

I think this approach provides us with the only sensible reading of Darwin's well-known remark that 'if it could be demonstrated that any complex organ existed, which could not possibly have been formed by numerous, successive, slight modifications, my theory would absolutely break down'. That is not a plea for gradualism, as a modern paleobiologist uses the term. Darwin's theory is falsifiable, but he was much too wise to make his theory *that* easy to falsify! Why on earth *should* Darwin have committed himself to such an arbitrarily restrictive version of evolution, a version that positively invites falsification? I think it is clear that he didn't. His use of the term 'complex' seems to me to be clinching. Gould describes this passage from Darwin as 'clearly invalid'. So it is invalid if the alternative to slight modifications is seen as DC-8 saltation. But if the alternative is seen as 747 saltation, Darwin's remark is valid and very wise. His theory is indeed falsifiable, and in the passage quoted he puts his finger on one way in which it might be falsified.

There are two kinds of imaginable saltation, then, DC-8 saltation and 747 saltation. DC-8 saltation is perfectly possible, undoubtedly happens in the laboratory and the farmyard, and may have made occasional contributions to evolution.* 747 saltation is statistically ruled out unless there is supernatural intervention. In Darwin's own time, proponents and opponents of saltation often had 747 saltation in mind, because they believed in – or were arguing against – divine intervention. Darwin was hostile to (747) saltation, because he correctly saw natural selection as an *alternative* to the miraculous as an explanation for adaptive complexity. Nowadays saltation either means punctuation (which isn't saltation at all) or DC-8 saltation, neither of which Darwin would have strong objections to in principle, merely doubts about the facts. In the modern context, therefore, I do not think Darwin should be labelled a strong gradualist. In the modern context, I suspect that he would be rather open-minded.

It is in the anti-747 sense that Darwin was a passionate gradualist, and it is in the same sense that we must all be gradualists, not just with respect to life on Earth, but with respect to life all over the universe. Gradualism in this sense is essentially synonymous with evolution. The sense in which we may be non-gradualists is a much less radical, although still quite interesting, sense. The theory of evolution by jerks has been hailed on television and elsewhere as radical and revolutionary, a paradigm shift. There is, indeed, an interpretation of it which is revolutionary, but that interpretation (the 747 macro-mutation version) is certainly wrong, and is apparently not held by its original proponents. The sense in which the theory might be right is

---

* As I was later to argue in 1989 when I coined the phrase 'evolution of evolvability' (in *Artificial Life*, a volume edited by Christopher Langton). I suggested there that, although rare, certain key stages in evolution, such as the origin of segmented body plans, may have occurred as sudden saltations. The first segmented animal could easily have had two segments. It would not have had one and a half.

not particularly revolutionary. In this field you may choose your jerks so as to be revolutionary, *or* so as to be correct, but not both.

Theory 5. Random evolution

Various members of this family of theories have been in vogue at various times. The 'mutationists' of the early part of the twentieth century – de Vries, W. Bateson and their colleagues – believed that selection served only to weed out deleterious freaks, and that the real driving force in evolution was mutation pressure. Unless you believe mutations are directed by some mysterious life force, it is sufficiently obvious that you can be a mutationist only if you forget about adaptive complexity – forget, in other words, most of the consequences of evolution that are of any interest! For historians there remains the baffling enigma of how such distinguished biologists as de Vries, W. Bateson and T. H. Morgan could rest satisfied with such a crassly inadequate theory. It is not enough to say that de Vries's view was blinkered by his working only on the evening primrose. He only had to look at the adaptive complexity in his own body to see that 'mutationism' was not just a wrong theory; it was an obvious non-starter.

These post-Darwinian mutationists were also saltationists and anti-gradualists, and Mayr treats them under that heading, but the aspect of their view that I am criticizing here is more fundamental. It appears that they actually thought that mutation, on its own without selection, was sufficient to explain evolution. This *could* not be so on any non-mystical view of mutation, whether gradualist or saltationist. If mutation is undirected, it is clearly unable to explain the adaptive directions of evolution. If mutation is directed in adaptive ways we are entitled to ask how this comes about. At least Lamarck's principle of use and disuse makes a valiant attempt at explaining how variation might be directed towards improvement. The 'mutationists' didn't even seem to see that there was a problem, possibly because they underrated the importance of adaptation – and they were not the last to do so. The irony with which we must now read W. Bateson's dismissal of Darwin is almost painful: 'The transformation of masses of populations by imperceptible steps guided by selection is, as most of us now see, so inapplicable to the fact that we can only marvel . . . at the want of penetration displayed by the advocates of such a proposition.'

Nowadays some population geneticists describe themselves as supporters of 'non-Darwinian evolution'. They believe that a substantial number of the gene replacements that occur in evolution are non-adaptive substitutions of alleles, whose effects are indifferent relative to one another. This may well be true. But it obviously has nothing whatever to contribute to solving the problem of the evolution of adaptive complexity. Modern advocates of neutralism admit that their theory cannot account for adaptation.

The phrase 'random genetic drift' is often associated with the name of Sewall Wright, but Wright's conception of the relationship between random drift and adaptation is altogether subtler than the others I have mentioned. Wright does not belong in Mayr's fifth category, for he clearly sees selection as the driving force of adaptive evolution. Random drift may make it easier for selection to do its job by assisting the escape from local optima, but it is still selection that is determining the rise of adaptive complexity.*

---

* Sewall Wright was the American member of the great triumvirate – the others were R. A. Fisher and J. B. S. Haldane – who founded population genetics and reconciled Darwinism with Mendelian genetics. Wright espoused random genetic drift in evolution. But he saw drift as a way in which, indirectly, adaptation might be improved. One of the problems with strong selection, as engineers know from their 'hill-climbing' algorithms, is getting trapped on local optima – hillocks within sight of an unreachable mountain. Wright's version of random drift enables a lineage to drift down the slope of a hillock into the valley, whereupon selection can then take over and propel it up the slopes of a much larger mountain. For Wright, drift alternating with selection allows greater perfection of adaptation than selection on its own. An excellent and brilliant suggestion.

Recently paleontologists have come up with fascinating results when they perform computer simulations of 'random phylogenies'. These random walks through evolutionary time produce trends that look uncannily like real ones, and it is disquietingly easy, and tempting, to read into the random phylogenies apparently adaptive trends which, however, are not there. But this does not mean that we can admit random drift as an explanation of real adaptive trends. What it might mean is that some of us have been too facile and gullible in what we think are adaptive trends. That does not alter the fact that there are some trends that really *are* adaptive – even if we don't always identify them correctly in practice – and those real adaptive trends can't be produced by random drift. They must be produced by some non-random force, presumably selection.

So, finally, we arrive at the sixth of Mayr's theories of evolution.

Theory 6. Direction (order) imposed on random variation by natural selection

Darwinism – the non-random selection of randomly varying replication entities by reason of their 'phenotypic' effects – is the only force I know that can, in principle, guide evolution in the direction of adaptive complexity. It works on this planet. It doesn't suffer from any of the drawbacks that beset the other five classes of theory, and there is no reason to doubt its efficacy throughout the universe.

The ingredients in a general recipe for Darwinian evolution are replicating entities of some kind, exerting phenotypic 'power' of some kind over their replication success. In *The Extended Phenotype* I referred to these necessary entities as 'active germ-line replicators' or 'optimons'. It is important to keep their replication conceptually separate from their phenotypic effects, even though, on some planets, there may be a blurring in practice. Phenotypic adaptations can be seen as tools of replicator propagation.

Gould disparages the replicator's-eye view of evolution as preoccupied with 'book-keeping'. The metaphor is a superficially happy one: it is easy to see the genetic changes that accompany evolution as book-keeping entries, mere accountant's records of the really interesting phenotypic events going on in the outside world. Deeper consideration, however, shows that the truth is almost the exact opposite. It is central and essential to Darwinian (as opposed to Lamarckian) evolution that there shall be causal arrows flowing from genotype to phenotype, but not in the reverse direction. Changes in gene frequencies are not passive book-keeping records of phenotypic changes: it is precisely because (and to the extent that) they actively *cause* phenotypic changes that evolution of the phenotype can occur. Serious errors flow, both from a failure to understand the importance of this one-way flow,* and from an over-interpretation of it as inflexible and undeviating 'genetic determinism'.

The universal perspective leads me to emphasize a distinction between what may be called 'one-off selection' and 'cumulative selection'. Order in the non-living world may result from processes that can be portrayed as a rudimentary kind of selection. The pebbles on a seashore become sorted by the waves, so that larger pebbles come to lie in layers separate from smaller ones. We can regard this as an example of the selection of a stable configuration out of initially more random disorder. The same can be said of the 'harmonious' orbital patterns of planets around stars, and electrons around nuclei, of the shapes of crystals, bubbles and droplets, even, perhaps, of the dimensionality of the universe in which we find ourselves. But this is all one-off selection. It does not give rise to progressive evolution because there is no replication, no succession

* The one-way flow from genotype to phenotype – from genes to bodies – becomes obvious when you contrast the effect of a gene mutation (bodies change in future generations) with a purely bodily 'mutation', such as when an animal loses a leg. The latter change does not go through to future generations. There is a one-way causal arrow from genes to body, which is not reversed. I am astonished that Gould, in his 'book-keeping metaphor', failed to see this. 'Book-keeping' profoundly and significantly misses the point.

of generations. Complex adaptation requires many generations of cumulative selection, each generation's change building upon what has gone before. In one-off selection, a stable state develops and is then maintained. It does not multiply, does not have offspring.

In life the selection that goes on *in any one generation* is one-off selection, analogous to the sorting of pebbles on a beach. The peculiar feature of life is that successive generations of such selection build up, progressively and cumulatively, structures that are eventually complex enough to foster the strong illusion of design. One-off selection is a commonplace of physics and cannot give rise to adaptive complexity. Cumulative selection is the hallmark of biology and is, I believe, the force underlying all adaptive complexity.

Other topics for a future science of Universal Darwinism

Active germ-line replicators together with their phenotypic consequences, then, constitute the general recipe for life; but the form of the system may vary greatly from planet to planet, both with respect to the replicating entities themselves, and with respect to the 'phenotypic' means by which they ensure their survival. Indeed, as Leslie Orgel has pointed out to me, the very distinction between 'genotype' and 'phenotype' may be blurred. The replicating entities do not have to be DNA or RNA. They do not have to be organic molecules at all. Even on this planet it is possible that DNA itself is a late usurper of the role, taking over from some earlier, inorganic crystalline replicator.*

A full science of Universal Darwinism might consider aspects of replicators transcending their detailed nature and the timescale

---

* This suggestion was persuasively made by the Scottish chemist Graham Cairns-Smith. I expounded his theory in *The Blind Watchmaker*, not because I necessarily believed it but because he so clearly emphasized the fundamental importance of replication in the origin of life.

over which they are copied. For instance, the extent to which they are 'particulate' as opposed to 'blending' probably has a more important bearing on evolution than their detailed molecular or physical nature. Similarly, a universe-wide classification of replicators might make more reference to their dimensionality and coding principles than to their size and structure. DNA is a digitally coded one-dimensional array. A 'genetic' code in the form of a two-dimensional matrix is conceivable. Even a three-dimensional code is imaginable, although students of Universal Darwinism will probably worry about how such a code could be 'read'. (DNA is, of course, a molecule whose three-dimensional structure determines how it is replicated and transcribed, but that doesn't make it a three-dimensional code. DNA's meaning depends upon the one-dimensional sequential arrangement of its symbols, not upon their three-dimensional position relative to one another in the cell.) There might also be theoretical problems with analogue, as opposed to digital codes, similar to the theoretical problems that would be raised by a purely analogue nervous system.*

As for the phenotypic levers of power by which replicators influence their survival, we are so used to their being bound up into discrete organisms or 'vehicles' that we forget the possibility of a more diffuse extra-corporeal or 'extended' phenotype. Even on this Earth a large amount of interesting adaptation can be interpreted as part of the extended phenotype. There is, however, a

* In nervous systems the problem is what engineers call random 'noise'. Some noise is added during any information transmission or amplification process. Because of the way neurones work, they are more susceptible to noise than, for example, telephone wires. Just as modern telephone systems have increasingly gone over to digital rather than analogue transmission, so neurones convey information by the temporal patterning of spikes rather than by the (analogue) height of the spikes. For a fuller discussion of analogue vs. digital, see the analogy with beacon fires and the Spanish Armada, in the essay entitled 'Science and sensibility' in section I of this collection (see pages 80–1).

general theoretical case that can be made in favour of the discrete organismal body, with its own recurrent life-cycle, as a necessity in any process of evolution of advanced adaptive complexity, and this topic might have a place in a full account of Universal Darwinism.

Another candidate for full discussion might be what I shall call divergence, and convergence or recombination, of replicator lineages. In the case of Earth-bound DNA, 'convergence' is provided by sex and related processes. Here the DNA 'converges' within the species after having very recently 'diverged'. But suggestions are now being made that a different kind of convergence can occur among lineages that originally diverged an exceedingly long time ago. For instance, there is evidence of gene transfer between fish and bacteria. The replicating lineages on other planets may permit very varied kinds of recombination, on very different timescales. On Earth, apart from in bacteria, the rivers of phylogeny are almost entirely divergent: if main tributaries ever recontact each other after branching apart it is only through the tiniest of trickling cross-streamlets, as in the fish/bacteria case. There is, of course, a richly anastomosing delta of divergence and convergence due to sexual recombination *within* the species, but only within the species. There may be planets on which the 'genetic' system permits much more cross-talk at all levels of the branching hierarchy, one huge fertile delta.

I have not thought enough about the fantasies of the previous paragraphs to evaluate their plausibility. My general point is that there is one limiting constraint upon all speculations about life in the universe. If a life form displays adaptive complexity, it must possess an evolutionary mechanism capable of generating adaptive complexity. However diverse evolutionary mechanisms may be, even if there is no other generalization that can be made about life all around the universe, I am betting it will always be recognizable as Darwinian life. The Darwinian Law may be as universal as the great laws of physics.

AFTERWORD

I promised in one of my footnotes in this piece to return to 'poetic science'. Steve Gould was so in love with his own rhetoric that he allowed his readers to confuse three kinds of discontinuity: macro-mutation, mass extinction and rapid gradualism. They have nothing else in common and the suggestion of a connection between them is worthless and profoundly misleading. Such is the danger of 'poetic science'. For the most extreme example I know of Gould's poetic rhetoric misleading even expert scientists, see the footnote to 'The "Alabama Insert"' on page 229.

I suspect that, on a less grand scale, poetic science bedevils medicine too. When my father, many years ago, developed a duodenal ulcer, the doctor told him he had to eat milk puddings and other mild, gentle foods. Later advice negates this. I suspect that the advice stemmed from the 'poetic' association of 'milky' with such attributes as 'mild', 'gentle' and 'soft' rather than from actual evidence. Poetic medicine! And today, if you want to lose weight you are encouraged not to eat butter, cream and other fatty foods. Again, is this advice based on evidence? Or is it in part at least a 'poetic' association with the word 'fat'?

I love the poetry of science in a good sense. That's why this book is called *Science in the Soul.* But there's bad poetry as well as good.

An ecology of replicators*

TODAY, NOTWITHSTANDING LOCAL school boards in various backwoods and boondocks of the United States, no educated person doubts the truth of evolution. Nor do they doubt the force of natural selection. Natural selection is not the only driver and guide of evolution. At least at the molecular level, random drift is also important; but selection is the only force capable of producing *adaptation*. When it comes to accounting for the stunning illusion of design in nature, there is no alternative to natural selection.[†] If a biologist denies the importance of natural selection in evolution, it is pretty safe to assume *not* that he has some alternative theory but

---

* Ernst Mayr was an illustrious German-American biologist, one of the founding fathers of the neo-Darwinian synthesis of the 1930s and 1940s. He could indeed be called the Grand Old Man of the synthesis, not least because he did live to be very old. He was a hundred when I met him and he was active and alert to the end. Among the many honours bestowed on him and the many publications devoted to his celebration was a *Festschrift* volume of the journal *Ludus Vitalis*, edited by the distinguished Spanish-American geneticist Francisco Ayala, to which I was invited to contribute the paper which appears here (slightly abbreviated). I dedicated it, 'with the very deepest respect, to Professor Ernst Mayr FRS, Hon. D.Sc. (Oxford) on the occasion of his hundredth birthday'.

† See the preceding essay in this collection, 'Universal Darwinism'.

that he simply underrates adaptation as a dominant property of life that needs explaining. Probably he has never set foot in a tropical rainforest, or set flipper over a coral reef, or set eyes on a David Attenborough film.

Nowadays, questions about adaptation are high in the consciousness of field biologists. It has not always been so. My old maestro Niko Tinbergen wrote of an experience when he was a young man: 'I still remember how perplexed I was upon being told off firmly by one of my zoology professors when I brought up the question of survival value after he had asked: "Has anyone an idea why so many birds flock more densely when they are attacked by a bird of prey?"' Today's student is more likely to be perplexed about what the professor could possibly have meant by his question if *not* survival value. People in Tinbergen's own field of ethology now complain of a stampeding backlash in the other direction, towards an overwhelming preoccupation with Darwinian survival value, at the expense of studies of behavioural mechanisms.

But still when I was learning biology at school, we were warned against a dire sin called 'teleology'. This was actually a warning against Aristotelian final causes, not against Darwinian survival value. Nevertheless, it perplexed me because I had never found final causes the slightest bit tempting. Any fool can see that a 'final cause' is not a cause at all.* It is just another name for the *problem* which, eventually, Darwin solved. Darwin showed how the illusion of a final cause could be produced by comprehensible efficient causes. His solution, refined by the giants of the modern synthesis

---

* Which makes it all the more surprising that Aristotle, no fool, could seriously entertain it. Aristotle is one of many highly intelligent thinkers who, one might suppose, could have worked out the principle of evolution by natural selection, but didn't. Why not? Evolution by natural selection is the kind of idea, you might think, that could occur to a great thinker and naturalist in any century. Unlike Newton's physics, it's hard to see why it needed two millennia of shoulders to stand on. However, it clearly did, so my intuition must be just plain wrong.

including Ernst Mayr, has put paid to biology's deepest mystery: the source of the illusion of design which pervades the living, but not the non-living world.

The illusion of design is at its strongest in the body shapes and behaviour patterns, the tissues and organs, the cells and molecules of individual creatures. The individuals of every species, without exception, show it powerfully. But there is another illusion of design which we notice at a higher level: the level of ecology. Design seems to reappear in the disposition of species themselves, in their arrangement into communities and ecosystems, in the dovetailing of species with species in the habitats which they share. There is a pattern in the intricate jigsaw of rainforest, say, or coral reef, which leads rhetoricians to preach disaster if but one component should be untimely ripp'd from the whole.

In extreme cases, such rhetoric takes on mystical tones. The planet is the womb of an earth goddess,* all life her body, the species her parts. Yet, without giving in to such hyperbole, there *is* a strong illusion of design at the community level, less compelling than within the individual organism but worth attention. The animals and plants that live together in an area seem to fit one another with something like the glove-like intimacy which the parts of an animal display as they mesh with other parts of the same organism.

A cheetah has the teeth of a carnivore, the claws of a carnivore, the eyes, ears, nose and brain of a carnivore, leg muscles that are suitable for chasing meat, and guts that are primed to digest it. Its parts are choreographed in a dance of carnivorous unity. Every

---

* Such mysticism reached its apogee in early versions of James Lovelock's Gaia hypothesis. In later versions Lovelock himself sought to disavow the mysticism, but it was still going strong at a conference where John Maynard Smith met a prominent devotee of 'Ecology' in the political rather than the scientific sense of the word. Somebody mentioned the theory that a large meteorite had struck Earth, thereby killing the dinosaurs. 'Of course not,' declared the fervent 'Ecologist', by Maynard Smith's account; 'Gaia would not have permitted it.'

sinew and cell of the big cat has meat-eater inscribed through its very texture, and we can be sure that this extends deep into the details of biochemistry. The corresponding parts of an antelope are equally unified with each other, but in pursuit of a different route to survival. Guts designed to digest plant roughage would be ill-served by claws and instincts designed to catch prey. And vice versa. A hybrid between a cheetah and an antelope would fall flat on its evolutionary face. Genetic tricks of the trade cannot be cut from one and pasted into the other. Their compatibility is with other tricks of the same trade.

Something similar can be said of communities of species. The language of the ecologist reflects this. Plants are primary producers. They trap energy from the sun, and make it available to the rest of the community, via a chain of primary, secondary and even tertiary consumers, culminating in scavengers. Scavengers play a recycling 'role' in the community. Every species, in this view of life, has a 'role' to play. In some cases, if the performers of some role, such as scavengers, were removed, the whole community would collapse. Or its 'balance' would be upset and it might fluctuate wildly, out of 'control' until a new balance is set up, perhaps with different species playing the same roles. Desert communities are different from rainforest communities and their component parts are ill-suited to other such communities just as – or so it seems – herbivorous colons are ill-suited to carnivorous teeth or hunting instincts. Coral-reef communities are different from sea-bottom communities, and their parts cannot be exchanged. Species become adapted to their community, not just to a particular physical region and climate. They become adapted to each other. The other species of the community are an important – perhaps the most important – feature of the environment to which each species becomes adapted. In one sense, the other species of the ecosystem are just another aspect of the weather. But unlike the temperature and the rainfall, the other species are themselves evolving. The illusion of design in ecosystems is an inadvertent consequence of this coevolution.

The harmonious role-playing of species in a community, then, resembles the harmony of the parts of a single individual organism. The resemblance is deceptive and must be treated with caution. We must not fall for the excesses of group-selectionist panglossianism such as the dubious concept of 'prudent predators'.* Given my biases, it feels like pulling teeth to say so, but the analogy between organism and community is not completely without foundation. It is one of my purposes in this article to argue that there is an ecology within the individual organism. I am not making the now commonplace point that a large metazoan body contains a community of bacteria, including mitochondria and other modified bacteria. I am making the much more radical suggestion that we should recognize the entire gene pool of a species as an ecological community of genes. The forces that produce harmony among the parts of an organism's body are not unlike the forces that produce the illusion of harmony in the species of a community. There is balance in a rainforest, structure in a reef community, an elegant meshing of parts whose coevolution recalls coadaptation within an animal body. In neither case is the balanced unit favoured *as a unit* by Darwinian selection. In both cases the balance comes about through selection at a lower level. Selection doesn't favour a harmonious whole. Instead, harmonious parts flourish in the presence of each other, and the illusion of a harmonious whole emerges.

At the individual level, to rehearse an earlier example in genetic language, genes that make carnivorous teeth will flourish in a gene pool containing genes that make carnivorous guts and carnivorous

---

* The ecologist Lawrence Slobodkin, who introduced the phrase, was later stung into indignant denial of the charge of group selectionism (*American Naturalist*, vol. 108, 1974). He may be right that a proper Darwinian defence of 'prudent predators' can – at a bit of a stretch – be mounted. But the phrase was ill-chosen. It begs to be interpreted along the lines of the Great Ecological Temptation – to forget the level at which natural selection actually works to produce individual adaptations, and think in terms of benefit to the group or even the community.

brains, but not in a gene pool containing genes for herbivorous guts and brains. At the community level, an area that lacks carnivorous species might experience something similar to a human economy's 'gap in the market'. Carnivorous species that enter the area find themselves flourishing. If the area is a remote island which no carnivorous species has reached, or if a recent mass extinction has devastated the land and created a similar gap in the market, natural selection will favour individuals within non-carnivorous species that change their habits and eventually their bodies, and become carnivores. After a long enough period of evolution, specialist carnivore species will be found to have descended from omnivorous or herbivorous ancestors.

Carnivores flourish in the presence of herbivores, and herbivores flourish in the presence of plants. But what about the other way around? Do plants flourish in the presence of herbivores? Do herbivores flourish in the presence of carnivores? Do animals and plants need enemies to eat them in order to flourish? Not in the straightforward way that is suggested by the rhetoric of some ecological activists. No creature benefits directly from being eaten. But grasses that can withstand being cropped better than rival plant species really do flourish in the presence of grazers – on the principle of 'my enemy's enemy'. And something like the same story might be told of the victims of parasites – and predators, although here it is more complicated. It is still misleading to say that a community 'needs' its parasites and predators like a polar bear needs its liver or its teeth. But the enemy's enemy principle does lead to something like the same result. It can be right to see a community of species as a kind of balanced entity which is potentially threatened by removal of any of its parts.

The idea of community, as made up of lower-level units that flourish in the presence of each other, pervades life. But, as I have said, I want to go beyond the familiar point that animal cells are communities of hundreds or thousands of bacteria. This is not to downplay the importance of bacterial symbioses. Mitochondria and chloroplasts have become so comprehensively integrated into the smooth workings of the cell that their bacterial origins have only recently become understood. Mitochondria are as essential to the workings of our cells as our cells are to them. Their genes have flourished in the presence of ours, as ours have flourished in the presence of theirs. Plant cells by themselves are incapable of photosynthesis. That chemical wizardry is performed by guest workers within the cells, originally bacteria and now relabelled chloroplasts. Plant eaters such as ruminants and termites are themselves largely incapable of digesting cellulose. But they are good at finding and chewing plants. The gap in the market offered by their plant-filled guts is exploited by symbiotic micro-organisms that possess the biochemical expertise necessary to digest plant material efficiently, and to the benefit of their herbivorous hosts. Creatures with complementary skills flourish in each other's presence.

My point is that the process is mirrored at the level of every species' 'own' genes. The entire genome of a polar bear or a penguin, of an iguana or a guanaco, is a set of genes that flourish in each other's presence. The immediate arena of this flourishing is the interior of an individual's cells. But the long-term arena is the gene pool of the species. Given sexual reproduction, the gene pool is the habitat of every gene as it is recopied and recombined down the generations.

This gives the species its singular status in the taxonomic hierarchy. Nobody knows how many separate species there are in the world, but, thanks largely to Ernst Mayr, we at least know what it would mean to count them. Arguments about whether there are thirty million separate species, as some have estimated, or only five million, are real arguments. The answer matters. Arguments about how many genera there are, or how many orders, families, classes or phyla, have no more status than arguments about how many tall men there are. It's up to you how you define tall, and it is up to you how you define a genus or a family. But – as long as reproduction is sexual – the species has a definition which goes beyond individual taste, and does so in a way that really matters. Fellow members of a species, by definition, can interbreed and therefore participate in the same shared gene pool. The species is defined as the community whose genes share that most intimate of arenas for cohabitation, the cell nucleus – a succession of cell nuclei through generations. When a species spawns a daughter species, usually after a period of accidental isolation, the new gene pool constitutes a new arena in which intergene cooperation can evolve. All the diversity on Earth has come about through such splittings. Every species is a unique entity, a unique set of coadapted genes, cooperating with each other in the enterprise of building individual organisms of that species.

The gene pool of a species is an edifice of harmonious coopera-tors, built up through a unique history. Any gene pool, as I have argued elsewhere, is a unique written record of ancestral history. Slightly fanciful perhaps, but it follows indirectly from Darwinian natural selection. A well-adapted animal reflects, in minute detail even down to the biochemical, the environments in which its ancestors survived. A gene pool is carved and whittled, through generations of ancestral natural selection, to fit that environment. In theory a knowledgeable zoologist, presented with the complete transcript of a genome, should be able to reconstruct the environ-mental circumstances that did the carving. In this sense the DNA is a coded description of ancestral environments, a 'genetic book of the dead'. George Williams said it in different words: 'A gene pool is an imperfect record of a running average of selection pressures over a long period of time in an area often much larger than individual dispersal distances.'

The gene pool of a species, then, is the rainforest in which the ecology of the genes flourishes. But why have I called my article an ecology of *replicators*? In answering this, I need to step back and look at a controversy in evolutionary theory, one in which Ernst Mayr has been an eloquent partisan. It is the controversy over the unit in the hierarchy of life at which natural selection may be said to act. In Richard Alexander's phrase, 'The fittest *what*?' Ernst Mayr and I have both coined a word – 'selecton' in his case, 'optimon' in mine – for the sole purpose of asking the *question*:

'What is the entity about which you may say that an adaptation is good for it?' Is it for the good of the group, the individual, the gene, life as a whole, or what? My own answer to the question – the gene – is not the answer Ernst Mayr would give – the organism. I shall try to show that the difference is apparent, not real. It will disappear when terminological differences are sorted out. After such a presumptuous – not to say impertinent – promise, let me try to deliver on it.

The wrong way to set up the debate is as a choice between rungs on a ladder, of which the gene is the lowest: gene, cell, organism, group, species, ecosystem. What is wrong with the ladder of levels is that the gene is really in a different category from all the rest. The gene is what I have called a replicator. All the rest are, if anything, 'vehicles' for replicators. The justification for treating the gene as special in this list of levels was clearly presented by Williams in 1966:

> The natural selection of phenotypes cannot in itself produce cumulative change, because phenotypes are extremely temporary manifestations. The same argument holds for genotypes ... Socrates' genes may be with us yet, but not his genotype, because meiosis and recombination destroy genotypes as surely as death ... It is only the meiotically dissociated fragments of the genotype that are transmitted in sexual reproduction, and these fragments are further fragmented by meiosis in the next generation. If there is an ultimate indivisible fragment it is, by definition, 'the gene' that is treated in the abstract discussions of population genetics.

Philosophers now call this 'genic selectionism', but I doubt that Williams regarded it as a radical departure from orthodox neo-Darwinian 'individual selection'. Nor did I, when I reiterated and extended the same argument a decade later in *The Selfish Gene* and *The Extended Phenotype*. We thought we were just clarifying what orthodox neo-Darwinism really meant. Yet both critics and supporters misunderstood our view as an attack on the Darwinian idea of the individual organism as the unit of selection. This was because we had not then made sufficiently clear the distinction between *replicators* and *vehicles*. Of course the individual organism is the unit (or at least a very important unit) of selection if you mean unit in the sense of vehicle. But it isn't a replicator at all.

A replicator is anything of which copies are made. An individual organism is not a replicator in this sense, and individual reproduction is not replication. Not even where it is asexual, clonal reproduction. This is not a matter of fact but a matter of definition. If you doubt it, you have not grasped the significance of the term 'replicator'.

For an operational criterion for whether an entity is a true replicator, ask what is the fate of *blemishes* in entities of its class. An individual organism, such as a clonally reproducing aphid or stick insect, would be a true replicator only if blemishes in the phenotype – say an amputated leg – were reproduced in the next generation. And of course they are not. Note that a blemish in the genotype – a mutation – *is* reproduced in the next generation. Of course it may then show itself in the phenotype too, but it is not the phenotypic blemish itself which is copied. This is no more than the familiar principle of the non-inheritance of acquired characteristics, or – its molecular version – Crick's Central Dogma.

I have described a replicator as 'active' if something about its nature affects its proficiency in being copied, which implies that blemished replicators may be less proficient, or more proficient, than the original (in practice because of what we are accustomed to calling 'phenotypic effects'). The true unit of selection in any Darwinian process, on any planet, is an active germ-line replicator. On this planet, it happens to be DNA.

Williams has returned to the matter in his more recent book, *Natural Selection*. He agrees that the gene does not belong in the same hierarchical list as the organism: 'These complications are best handled by regarding individual selection, not as a level of selection in addition to that of the gene, but as the primary mechanism of selection at the genic level.'

'Primary mechanism of selection at the genic level' is Williams' way of describing what I would call the 'vehicle', and David Hull would call the 'interactor'. Williams' version of my 'replicator' – in other words, his way of singling out the gene from all vehicles – is to coin the phrase *codical domain*, as opposed to *material domain*. A member of the codical domain is a codex. The information coded in a gene is firmly in the codical domain. The atoms in the DNA of the gene are in the material domain. The only other candidates I can think of for the codical domain are memes such as self-replicating computer programs, and units of cultural inheritance. Which is to say that both of these are candidates for the title of active germ-line replicator, and candidates for the basic unit of selection in a hypothetical Darwinian process. The individual organism is not even a candidate for a replicator in any Darwinian process, however hypothetical.

But I haven't yet done justice to the criticisms of the idea of genic selectionism. The most cogent of these criticisms have come from Ernst Mayr himself, using arguments foreshadowed in his famous attack on what he provocatively* called beanbag genetics, and in the 'Unity of the genotype' chapter of *Animal Species and Evolution*. In that chapter, for example, he said: 'To consider genes as independent units is meaningless from the physiological as well as from the evolutionary viewpoint.'

This beautifully written book is a favourite of mine, and I agree with every word of the 'Unity of the genotype' chapter except its take-home message, which I disagree with profoundly!

The important thing is to distinguish between the role of genes in embryology and the role of genes in evolution. It is undeniably the case – but completely irrelevant to the levels of selection debate – that genes interact with each other in inextricably complex ways in embryology, even if not all embryologists would go so far

---

* It provoked J. B. S. Haldane's spirited 'Defence of beanbag genetics'. Beanbag genetics, in this context, means the quantitative treatment of changes in gene frequency in populations, treating genes as particulate Mendelian entities.

as Mayr in saying: 'Every character of an organism is affected by all genes and every gene affects all characters.'

Mayr himself acknowledges that this was exaggerated. I am happy to quote it in the same spirit. Happy to quote it because, even if it were literally true, it would not undermine, not even to the tiniest extent, the status of the gene as the unit of selection: unit in the sense of replicator, that is. If this sounds like a paradox, the resolution is actually given by Mayr himself: 'A given gene has as its genetic environment not only the genetic background of the given zygote on which it is temporarily placed, but the entire gene pool of the local population in which it occurs.'

This really is the key point. Every gene is selected for its ability to survive in its environment. We naturally think first of the external environment. But the most important elements in the environment of a gene are the other genes. This 'ecology of genes', in which each is separately selected for its ability to flourish in the presence of the others in the sexually recombining gene pool, is what creates the illusion of 'unity of the genotype'. It is emphatically not right to say that because the genome is unified in its embryological role, it is therefore also unified in its evolutionary role. Mayr was right about embryology. Williams was right about evolution. There is no disagreement.

Twelve misunderstandings
of kin selection*

Introduction

KIN SELECTION HAS become a bandwagon, and when
bandwagons start to roll attitudes sometimes polarize. The
rush to jump on provokes a healthy reaction. So it is that today the

* The theory of kin selection – natural selection favours genes for
helping relatives because they are statistically likely to be present in
the relatives helped – was developed by W. D. Hamilton, who later
became my Oxford colleague and friend. It was one of the central
themes of my first book, *The Selfish Gene*. Having been largely
neglected for its first decade after Hamilton's important papers of
1964, the theory of kin selection suddenly became much discussed
in the mid-1970s by biologists and the wider world. Kin selection's
popularity spawned a rich plethora of misunderstandings, some of
the more bizarre ones being perpetrated by distinguished social
scientists who – dare one suggest – may have felt threatened by this
sudden incursion into what they thought was their field. This
upsurge of wayward commentary prompted me to collect twelve of
these misunderstandings and refute them in an article published
(in English) in the leading German journal of animal behaviour,
the *Zeitschrift für Tierpsychologie*. As usual in a scientific paper,
there were many references to the literature. These have been
omitted here. I have also cut three of the misunderstandings,
numbers 8, 9 and 11. Although they are important, they concern
technicalities which could only be made clear by providing an
excess of space-filling background information.

sensitive ethologist with his* ear to the ground detects a murmuring of sceptical growls, rising to an occasional crescendo of smug baying when one of the early triumphs of the theory encounters new problems. Such polarization is a pity. In this case it is exacerbated by a notable series of misunderstandings, both on and off the kin selection bandwagon. Many of these misunderstandings arise from secondary attempts at explaining Hamilton's ideas rather than from his original mathematical formulation. As one who has fallen for some of them in my time and met all of them frequently, I would like to try the difficult exercise of explaining in non-mathematical language twelve of the commonest misunderstandings of kin selection. The twelve by no means exhaust the supply. Alan Grafen, for instance, has published good exposés of two other, rather more subtle ones. The twelve sections can be read in any order.

Misunderstanding 1: 'Kin selection is a special, complex kind of natural selection, to be invoked only when "individual selection" proves inadequate'

This one logical error, on its own, is responsible for a large part of the sceptical backlash that I mentioned. It results from a confusion between historical precedence and theoretical parsimony: 'Kin selection is a recent addition to our theoretical armoury; for many purposes we got along quite well without it for years; therefore we should turn to it only when good old-fashioned "individual selection" fails us.'

Note that good old-fashioned individual selection has always included parental care as an obvious consequence of selection for

---

* Today my consciousness has been raised to the point where I'd say 'with her ear to the ground'. Not 'his or her', which sounds obtrusively clumsy to *my* ear. I prefer the convention by which authors signal courteous respect for the sex opposite to their own by favouring the appropriate pronouns. Ethology is the biological study of animal behaviour. Nowadays I could equally well have said 'sensitive sociobiologist', 'sensitive behavioural ecologist' or 'sensitive evolutionary psychologist' with her ear to the ground.

individual fitness. What the theory of kin selection has added is that parental care is only a special case of caring for close relatives. If we look in detail at the genetical basis of natural selection, we see that 'individual selection' is anything but parsimonious, while kin selection is a simple and inevitable consequence of the differential gene survival that, fundamentally, *is* natural selection. Caring for close relatives at the expense of distant relatives is predicted from the fact that close relatives have a high chance of propagating the gene or genes 'for' such caring: the gene cares for copies of itself. Caring for oneself and one's own children *but not* equally close collateral relatives is hard to predict by any simple genetic model. We have to invoke additional factors, such as the assumption that offspring are easier to identify or easier to help than collateral relatives. These additional factors are perfectly plausible but they have to be *added* to the basic theory.

It happens to be true that most animals care for offspring more than they care for siblings, and it is certainly true that evolutionary theorists understood parental care before they understood sibling care. But neither of these two facts implies that the general theory of kin selection is unparsimonious. If you accept the genetical theory of natural selection, as all serious biologists now do, then you must accept the principles of kin selection. Rational scepticism is limited to beliefs (perfectly sensible ones) that *in practice* the selection pressure in favour of caring for relatives other than offspring is unlikely to have noticeable evolutionary consequences.*

Misunderstanding 1 has perhaps been unwittingly encouraged by an influential definition of kin selection propagated by Edward

---

* 'Hamilton's Rule' succinctly sums up his theory. A gene for altruism will spread through the gene pool if $rB > C$, i.e. if the cost C to the altruist is outweighed by the benefit B to the recipient multiplied by a fraction r, representing the closeness of genetic relationship between them. The reason parental care is commoner than care of full siblings is that, although r is the same for both relationships (0.5), the B and C terms in practice favour parental care.

O. Wilson: 'the selection of genes due to one or more individuals favoring or disfavoring the survival and reproduction of relatives (other than offspring) who possess the same genes by common descent'. I am glad to see that Wilson has omitted the phrase 'other than offspring' in his more recent definition, in favour of the following: 'Although kin are defined so as to include offspring, the term kin selection is ordinarily used only if at least some other relatives, such as brothers, sisters, or parents, are also affected'. This is undeniably true, but I still think it is regrettable. Why should we treat parental care as special, just because for a long time it was the only kind of kin-selected altruism we understood? We do not separate Neptune and Uranus off from the rest of the planets simply because for centuries we did not know of their existence. We call them all planets because they are all the same kind of thing.

At the end of his 1975 definition, Wilson added that kin selection was 'one of the extreme forms of group selection'. This, too, has happily been deleted from his 1978 definition.* It is the second of my twelve misunderstandings.

Misunderstanding 2: 'Kin selection is a form of group selection'

Group selection is the differential survival or extinction of whole groups of organisms. It happens that organisms sometimes go around in family groups, and it follows that differential group extinction could turn out to be effectively equivalent to family selection or 'kin group selection'. But this has nothing to do with the essence of Hamilton's basic theory: those *genes* are selected that tend to make individuals discriminate in favour of other individuals who are especially likely to contain copies of the same genes. The population does not need to be divided up into family groups in

---

* Unfortunately both these improvements have been reversed by Wilson in more recent publications, including his book *The Social Conquest of Earth*, in ways that suggest to me that he never really understood kin selection in the first place.

order for this to happen, and it is certainly not necessary that whole families should go extinct or survive as units.

Animals cannot, of course, be expected to know, in a cognitive sense, who their relatives are (see Misunderstanding 3), and in practice the behaviour that is favoured by natural selection will be equivalent to a rough rule of thumb such as 'share food with anything that moves, in the nest in which you are sitting'. If families happen to go around in groups, this fact provides a useful rule of thumb for kin selection: 'Care for any individual you often see.' But note once again that this has nothing to do with true group selection: differential survival and extinction of whole groups do not enter into the reasoning. The rule of thumb works if there is any 'viscosity' in the population such that individuals are statistically likely to encounter relatives; there is no need for families to go about in discrete groups.

Hamilton is perhaps right to blame the phrase 'kin selection' itself for some misunderstanding, ironically since it was coined (by Maynard Smith) with the laudable purpose of emphasizing its distinctness from group selection. Hamilton himself does not use the phrase, preferring to stress the relevance of his central concept of inclusive fitness* to any kind of genetically non-random altruism, whether concerned with kin-relatedness or not. For instance, suppose that within a species there is genetic variation in habitat choice. Suppose further that one of the genes contributing to this variation has the pleiotropic† effect of making individuals share food with others whom they encounter. Because of the pleiotropic

---

* Hamilton gave 'inclusive fitness' a more precise mathematical definition which can be rendered, somewhat lengthily, into words, but he approved my informal definition: 'Inclusive fitness is that quantity which an individual will appear to be maximizing when what is really being maximized is the survival of its genes.'

† Many genes have more than one effect, often apparently unconnected with each other, and the phenomenon is called pleiotropy.

effect on habitat choice, this altruistic gene is effectively discriminating in favour of copies of itself, since individuals possessing it are especially likely to congregate in the same habitat and therefore meet each other. They do not have to be close kin.

Any way in which an altruistic gene can 'recognize' copies of itself in other individuals could form the basis for a similar model. The principle is reduced to its bare essentials in the improbable but instructive 'Green Beard Effect': selection would theoretically favour a gene that pleiotropically caused individuals to develop a green beard and also a tendency to be altruistic to green-bearded individuals. Again there is no need for the individuals to be kin.*

Misunderstanding 3: 'The theory of kin selection demands formidable feats of cognitive reasoning by animals'

In a much-quoted 'anthropological critique of sociobiology', Sahlins† says the following:

> In passing it needs to be remarked that the epistemological problems presented by a lack of linguistic support for calculating r, coefficients of relationship, amount to a serious defect in the theory of kin selection. Fractions are of very rare occurrence in the world's languages, appearing in Indo-European and in the archaic

---

* The Green Beard Effect is an unrealistic hypothetical, a parable. What is realistic – and this is the point of the parable – is that kinship acts as a kind of statistical green beard. An animal with a genetic propensity to care for full siblings, say, has a 50 per cent chance that it is caring for copies of itself. Brotherhood is a label like a green beard. We do not expect animals to be cognitively aware of brotherhood. In practice the label is likely to be something like 'he who sits in the same nest as you'.

† Marshall Sahlins is a distinguished American anthropologist. Some other anthropologists have taken the trouble to learn some biology. To be fair, I expect I would exhibit similar ignorance and lack of understanding, were I to wade into the field of anthropology. But I do not so wade.

civilizations of the Near and Far East, but they are generally lacking among the so-called primitive peoples. Hunters and gatherers generally do not have counting systems beyond *one*, *two* and *three*. I refrain from comment on the even greater problem of how animals are supposed to figure out how that r [ego, first cousins] = 1/8. The failure of sociobiologists to address this problem introduces a considerable mysticism in their theory.

A pity, for Sahlins, that he succumbed to the temptation to 'refrain from comment' on how 'animals are supposed to figure out' r. The very absurdity of the idea he tried to ridicule should have set mental alarm bells ringing. A snail shell is an exquisite logarithmic spiral, but where does the snail keep its log tables; how indeed does it read them, since the lens in its eye lacks 'linguistic support' for calculating μ, the coefficient of refraction? How do green plants 'figure out' the formula of chlorophyll? Enough, let us be constructive.

Natural selection chooses genes rather than their alleles,* because of those genes' phenotypic effects. In the case of behaviour, the genes presumably influence the state of the nervous system, which in turn influences the behaviour. Whether it is behaviour, physiology or anatomy, a complex phenotype may require sophisticated mathematical description if we are to understand it. This does not, of course, mean that the animals themselves have to be mathematicians. Unconscious 'rules of thumb' of the kind already mentioned will be selected. For a spider to build a web, rules of thumb are required that are probably more sophisticated than any that kin-selection theorists have postulated. If spider webs did not exist, anybody who postulated them might well provoke scornful

---

* 'Alleles' are alternative forms of a gene which vie for a particular slot or 'locus' on a chromosome. In sexually reproducing creatures, natural selection can be seen as competition between alleles in the gene pool for that slot. The weapons of their competition are normally the 'phenotypic' effects that they have on bodies.

scepticism. But they do exist; we have all seen them, and nobody wonders how spiders 'figure out' the designs.

The machinery that automatically and unconsciously builds webs must have evolved by natural selection. Natural selection means the differential survival of alleles in gene pools. There must, therefore, have been genetic variation in the tendency to build webs. Similarly, to talk about the evolution of altruism by kin selection we have to postulate genetic variation in altruism. In this sense we have to postulate alleles 'for' altruism, to compare with alleles for selfishness. This brings me to my next misunderstanding.

Misunderstanding 4: 'It is hard to imagine a gene "for" anything so complex as altruistic behaviour towards kin'

The problem results from a misunderstanding about what it means to speak of a gene 'for' behaviour. No geneticist has ever imagined that a gene 'for' some phenotypic character such as microcephaly, or brown eyes, is responsible, alone and unaided, for the manufacture of the organ that it affects. A microcephalic head is abnormally small, but it is still a head, and a head is much too complex a thing to be made by a single gene. Genes don't work in isolation, they work in concert. The genome as a whole works with its environment to produce the body as a whole.

Similarly, 'a gene for behaviour X' can only refer to a *difference* between the behaviour of two individuals. Fortunately, it is precisely such differences between individuals that matter for natural selection. When we speak of the natural selection of, for instance, altruism towards younger siblings, we are talking of the differential survival of a gene or genes 'for' sibling altruism. But this simply means a gene that tends to make individuals in a normal environment more likely to show sibling altruism than they would under the influence of an allele of that gene. Is that implausible?

It is true that no geneticist has actually bothered to study genes for altruism. Nor has any geneticist studied web-building in spiders. We all believe that web-building in spiders has evolved under the influence of natural selection. This can only have happened if, at each and every step of the evolutionary way, genes for some difference in spider behaviour were favoured over their alleles. This does not, of course, mean there still have to be such genetic differences; natural selection could, by now, have removed the original genetic variance.

Nobody denies the existence of maternal care, and we all accept that it has evolved under the influence of natural selection. Again, we don't need to do genetic analysis to convince ourselves that this can only have happened if there were a series of genes for various behaviour differences which, together, built up maternal behaviour. Once maternal behaviour, in all its complexity, exists, it takes little imagination to see that only a small genetic change is required to push it over into elder sibling altruism.

Suppose the 'rule of thumb' that mediates maternal care in a bird is the following: 'Feed anything that squawks inside your nest.' This is plausible, since cuckoos seem to have exploited some such simple rule. Now all that is needed to obtain sibling altruism is a slight quantitative shift, perhaps a small postponement of a fledgling's departure from the parental nest. If it postpones its departure until after the next brood has hatched, its existing rule of thumb might well cause it automatically to start feeding the squawking gapes that have suddenly appeared in its home nest. Such a slight quantitative postponing of a life-historical event is exactly the kind of thing a gene can be expected to effect. In any case the shift is child's play compared with those that must have accumulated in the evolution of maternal care, web-building, or any other undisputed complex adaptation. Misunderstanding 4 turns out to be only a new version of one of the oldest objections to Darwinism itself, an objection that Darwin anticipated and decisively disposed of in his section of the *Origin* on 'Organs of extreme perfection and complication'.

Altruistic behaviour may be very complex, but it got its complexity, not from a new mutant gene, but from the pre-existing developmental process that the gene acted upon. There already was complex behaviour before the new gene came along, and that complex behaviour was the result of a long and intricate developmental process involving a large number of genes and environmental factors. The new gene of interest simply gave this existing complex process a crude kick, the end result of which was a crucial change in the complex phenotypic effect. What had been complex maternal care, say, became complex sibling care. The shift from maternal to sibling care was a simple one, even if both maternal and sibling care are very complex in themselves.

Misunderstanding 5: 'All members of a species share more than 99 per cent of their genes, so why shouldn't selection favour universal altruism?'

> This whole calculus upon which sociobiology is based is grossly misleading. A parent does not share one half of the genes with its offspring; the offspring shares one half of the genes in which the parents differ. If the parents are homozygous for a gene, obviously all offspring will inherit that gene. The issue then becomes: How many shared genes are there within a species such as *Homo sapiens*? King and Wilson estimate that man and chimpanzee share 99 per cent of their genetic material; they also estimate that the races of man are 50 times closer than are man and chimpanzee. Individuals whom sociobiologists consider unrelated share, in fact, more than 99 per cent of their genes. It would be easy to make a model in which the structure and physiology important in behavior are based on the shared 99 per cent and in which behaviorally unimportant differences, such as hair form, are determined by the 1 per cent. The point is that genetics actually supports the beliefs of the social sciences, not the calculations of the sociobiologists.

This misconception, by another distinguished anthropologist, Sherwood Washburn, arises not from Hamilton's own mathematical formulation but from oversimplified secondary sources to which Washburn refers. The mathematics, however, are difficult, and it is worth trying to find a simple verbal way of refuting the error.

Whether 99 per cent is an exaggeration or not, Washburn is certainly right that any two random members of a species share the great majority of their genes. What, then, are we talking about when we speak of the coefficient of relatedness r between, say, siblings as being 50 per cent? We must answer this question first before getting down to the error itself.

The unqualified statement that parents and offspring share 50 per cent of their genes is, as Washburn rightly says, false. It can be made true by means of a qualification. A lazy way of qualifying it is to announce that we are only talking about rare genes; if I have a gene that is very rare in the population as a whole, the probability that my child or my brother has it is about 50 per cent. This is lazy because it evades the important fact that Hamilton's reasoning applies at all frequencies of the gene in question; it is an error (see Misunderstanding 6) to suppose that the theory only works for rare genes. Hamilton's own way of qualifying the statement is different. It is to add the phrase 'identical by descent'. Siblings may share 99 per cent of their genes altogether, but only 50 per cent of their genes are identical by descent, that is, are descended from the same copy of the gene in their most recent common ancestor.

So, we have identified two ways of explaining the meaning of r, the coefficient of relatedness: the 'rare gene' way and the 'identical by descent' way.* Neither of these, however, shows us how to escape from Washburn's paradox. Why is it not the case that natural selection will favour universal altruism, since most genes are universally shared in a species?

Let there be two strategies, Universal Altruist U, and Kin Altruist K. U individuals care for any member of the species indiscriminately. K individuals care for close kin only. In both cases, the caring behaviour costs the altruist something in terms of his personal survival chances. Suppose we grant Washburn's assumption that U behaviour 'is based on the shared 99 per cent of genes'. In other words, virtually the entire population are universal

---

* See also the footnote on pages 57–8.

altruists, and a tiny minority of mutants or immigrants are kin altruists. Superficially, the U gene appears to be caring for copies of itself, since the beneficiaries of its indiscriminate altruism are almost bound to contain the same gene. But is it evolutionarily stable against invasion by initially rare K genes?*

No, it is not. Every time a rare K individual behaves altruistically, it is especially likely to benefit another K individual *rather than* a U individual. U individuals, on the other hand, give out altruism to K individuals and U individuals indiscriminately, since the defining characteristic of U behaviour is that it is indiscriminate. Therefore K genes are bound to spread through the population at the expense of U genes. Universal altruism is not evolutionarily stable against kin altruism. Even if we assume it to be initially common, it will not remain common. This leads directly into the next, complementary, misunderstanding.

Misunderstanding 6: 'Kin selection only works for rare genes'

The logical outcome of the statement that, say, sibling altruism is favoured by natural selection, is that the relevant genes will spread to fixation.† Virtually all individuals in the population will be sibling

* 'Evolutionarily stable strategy' or ESS is John Maynard Smith's phrase, and it represents a powerful way of thinking about evolution, one of which I made extensive use in *The Selfish Gene*. A 'strategy' is a piece of unconscious behavioural 'clockwork' such as 'drop food into squawking gapes that you see in your nest'. An ESS is a strategy such that, when a majority of the population adopts it, it cannot be bettered by an alternative strategy. If it can be bettered, it is 'unstable'. A population dominated by an unstable strategy will be 'invaded' by the superior alternative strategy. ESS reasoning commonly starts with a statement such as: 'Imagine a strategy P, such that all members of the population are doing P. Now imagine that a new strategy, Q, were to arise by mutation; would natural selection cause Q to "invade"?' That is what we are doing in our reasoning about the strategies U and K.

† 'Fixation' is the technical term used by population geneticists

altruists. Therefore, if they did but know it, they would benefit the gene for sibling altruism just as much by caring for a random member of the species as by caring for a sibling! So it might seem that genes for exclusive kin altruism are favoured only when rare.

To put it this way is to expect animals, even genes, to play God. Natural selection is more mechanical than that.* The kin altruism gene does not program individuals to take intelligent action on its behalf; it specifies a simple behavioural rule of thumb such as 'feed squawking gapes in the nest in which you live'. It is this unconscious rule that will become universal when the gene becomes universal.

As in the case of the previous fallacy, we can use the language of evolutionarily stable strategies. We now ask whether kin altruism, K, is stable against invasion by universal altruism, U. That is, we assume that kin altruism has become common and ask whether mutant universal altruist genes will invade. The answer is no, for the same reason as before. The rare universal altruists care for the rival K allele indiscriminately with copies of their own U allele. The K allele, on the contrary, is especially unlikely to care for copies of its rival.

We have shown, therefore, that kin altruism is stable against invasion by universal altruism, but that universal altruism is not stable against invasion by kin altruism. This is the nearest I can get to a verbal explanation of Hamilton's mathematical argument that altruism to close relatives is favoured over universal altruism at all frequencies of the genes concerned. Although it lacks the mathematical precision of Hamilton's own presentation, it should at least suffice to remove these two common qualitative misunderstandings.

---

for the spread of a gene through the population until everybody, or almost everybody, has it. A gene can spread to fixation either (the interesting reason) because of positive natural selection, or through random chance, so-called 'genetic drift'.

* That's why I used the word 'clockwork' in my previous footnote defining ESS.

Misunderstanding 7: 'Altruism is necessarily expected between members of an identical clone'

There are races of parthenogenetic* lizards the members of which appear to be identical descendants, in each case, of a single mutant. The coefficient of relatedness between individuals within such a clone is 1. A naive application of rote-learned kin selection theory might therefore predict great feats of altruism between all members of the race. Like the previous one, this fallacy is tantamount to a belief that genes are god-like.

Genes for kin altruism spread because they are especially likely to help copies of themselves rather than of their alleles. But the members of a lizard clone all contain the genes of their original founding matriarch. She was part of an ordinary sexual population, and there is no reason to suppose that she had any special genes for altruism. When she founded her asexual clone, her existing genome was 'frozen': a genome that had been shaped by whatever selection pressures had been at work before the clonal mutation.

Should any new mutation for more indiscriminate altruism arise within the clone, the possessors of it would be, by definition, members of a new clone. Evolution could therefore, in theory, now occur by inter-clonal selection. But the new mutation would have to work via a new rule of thumb. If the new rule of thumb is so indiscriminate that both sub-clones benefit, the altruistic sub-clone is bound to decrease, since it is paying the cost of the altruism. We could imagine a new rule of thumb that initially achieved discrimination in favour of the altruistic sub-clone. But this would have to be something like an ordinary 'close-kin' altruism rule of thumb (e.g. 'care for occupants of your own nest'). Then if the sub-clone possessing this rule of thumb did indeed spread at the expense of the selfish sub-clone, what would we eventually see? Simply a race of lizards in which each one cared for occupants of her own

---

* *Parthenos* is Greek for 'virgin'. Parthenogenetic lizards reproduce without benefit of males, producing 'clonal' daughters equivalent to identical twins of themselves.

nest: not clone-wide altruism but ordinary 'close-kin' altruism. (Pedants please refrain from commenting that lizards don't have nests!)

I hasten to add, however, that there are other circumstances in which clonal reproduction is expected to lead to special altruism. Nine-banded armadillos have become a favourite talking point, because they reproduce sexually but each litter consists of four identical quadruplets. Here within-clone altruism is indeed expected, because genes are reassorted sexually in each generation in the usual way. This means that any gene for clonal altruism is likely to be shared by all members of some clones and no members of rival clones.

There is, so far, no good evidence for or against the predicted within-clone altruism in armadillos. However, some intriguing evidence in a comparable case has been reported by Aoki. In the Japanese aphid *Colophina clematis*, sisterhoods of asexually produced females consist of two types of individuals. Type A females are normal plant-sucking aphids. Type B females do not progress beyond 1st instar* and never reproduce. They have an abnormally short rostrum which is ill-adapted to sucking plants, and enlarged 'pseudoscorpion-like' prothoracic and mesothoracic legs. Aoki showed that Type B females attack large insects and kill them. He speculated that they constitute a sterile 'soldier caste' who protect their reproductive sisters against predators. It is not known how the 'soldiers' feed. Aoki doubts that their fighting mouthparts are capable of absorbing sap. He does not suggest that they are fed by their Type A sisters, but that fascinating possibility is presumably

---

* 'Instar' is the technical term used by entomologists for the discrete developmental stages that insects go through as they grow up. They're discrete and discontinuous because the insect skeleton consists not of internal bones, like ours, but of external armour. Unlike bone, the external armour, once hardened, can't grow; so the insect has to shed it periodically, then swell and grow a new suit of armour the next size up. Each of these incremental stages is an 'instar'.

open. He reports indications of similar soldier castes in other aphid genera.

There is a nice irony in Aoki's discussion, brought to my attention by R. L. Trivers. 'It may be concluded from [Hamilton's] theory that true sociality should occur more frequently in groups with haplodiploidy than in those without it . . . I do not know how many occurrences of true sociality among animals without haplodiploidy would be sufficient to refute his theory. The existence of soldiers in aphids should take part in one of the gravest problems against his theory, however.'*

* Aoki's spectacular error stems, like those of Sahlins and Washburn, from imperfect understanding of Hamilton's theory. Hamilton included in his exposition a brief section on 'haplodiploidy', the peculiar genetic system of the Hymenoptera – ants, bees and wasps. Females are diploid, like us, having chromosomes in pairs. Males, however, are haploid. They have half as many chromosomes as females. All the sperms produced by an individual male are therefore identical. Hamilton ingeniously pointed out a revealing consequence: the relatedness r between full sisters is 0.75 instead of the usual 0.5, since the paternal complement of their genes is identical. A female ant is more closely related to her full sister than to her daughter! This, as Hamilton pointed out, could predispose the Hymenoptera to supreme feats of social cooperation. This idea is so clever, so *charismatic* even, that many readers thought it was the whole point of his theory, instead of a throwaway couple of paragraphs – the icing on the cake. Aoki evidently was one such reader. If he had understood the whole gene-selection basis of Hamilton's theory, rather than just the few charismatic paragraphs, he would never have made his lamentable howler about his altruistic aphids. He thought they constituted a 'grave problem' for Hamilton's theory. In fact, given the right conditions, Hamilton's theory would predict even greater feats of social cooperation among clonal aphids than among ants, bees and wasps. The relatedness, r, between Aoki's aphids is 1.0 rather than the mere 0.75 of Hymenopteran sisters. Termites, by the way, are not haplodiploid, but Hamilton had a different ingenious idea for them, based on

This error is most instructive. *Colophina clematis*, like other aphids, have winged sexually reproducing dispersal phases interspersed with viviparous parthenogenetic generations. The 'soldiers' and the Type A individuals whom they seem to protect are wingless, and are almost certainly members of the same clone. The regular intervention of winged sexual generations ensures that genes for facultatively developing into a soldier, and alleles for not doing so, would be shuffled throughout the population. Some clones would therefore have such genes while rival clones would not. Conditions, in fact, are quite different from those of the lizards, and are ideal for the evolution of sterile castes. The soldiers and their reproductive clone mates are best regarded as parts of the same extended body. If a soldier aphid altruistically sacrifices her own reproduction, then so does my big toe. In almost exactly the same sense!

Misunderstanding 10: 'Individuals should tend to inbreed, simply because this brings extra close relatives into the world'

I have to be careful here, because there is a correct line of reasoning that sounds very like the error. Moreover, there may be other selection pressures for and against inbreeding, but these have nothing to do with the present argument: the proponent of the misconception is assumed to have covered himself with an 'other things being equal'.

The reasoning I wish to criticize runs as follows. Assume a monogamous mating system. A female who mates with a random male brings into the world a child related to her by $r = \frac{1}{2}$. If only she had mated with her brother she would have brought into the world a 'super-child' with an effective coefficient of relatedness of ¾. Therefore genes for inbreeding are propagated at the expense of

---

inbreeding, to explain their social cooperation. Not that special ingenuity is really necessary. There are many combinations of B and C which could combine with an r of 0.5 to promote social cooperation and even worker sterility.

genes for outbreeding, having a greater probability of getting into each child born.

The error is a simple one. If the individual refrains from mating with her brother, he is free to mate with some other female. So an outbreeding female gains a nephew/niece ($r = \frac{1}{4}$) plus a normal child of her own ($r = \frac{1}{2}$) to match the single super-child of the incestuous female (effective $r = \frac{3}{4}$). It is important to note that the refutation of the error assumes the equivalent of monogamy. If the species is, say, polygynous* with high variance in male reproductive success and a large bachelor population, things can be very different. It is now no longer true that a female, by mating with her brother, deprives him of the chance to mate with someone else. Most probably, the free mating his sister gives him is the only mating he will get. The female therefore does not deprive herself of an independent niece/nephew by mating incestuously, and she does bring into the world a child who is a super-child from her own genetic point of view. There may, then, be selection pressures in favour of incest, but the heading to this section is, as a general statement, incorrect.

Misunderstanding 12: 'An animal is expected to dole out to each relative an amount of altruism proportional to the coefficient of relatedness'

As S. Altmann has pointed out, I perpetrated this error when I wrote that 'second cousins should tend to receive one-sixteenth as much altruism as offspring or siblings.'[†] To oversimplify Altmann's argument, suppose I have a cake that I am going to give to my relatives, how should I divide it? The fallacy under discussion amounts to cutting the cake in such a way that each relative gets a

---

* One male, many females: harem-style reproduction. It's much more common than the reverse, polyandry, for reasons that are interesting but not necessary to spell out here.

† I should have said, 'Other things being equal, siblings are 16 times more likely to receive altruism than second cousins.'

slice whose size is proportional to his coefficient of relatedness to me. Really, of course, there is better reason to give the entire cake to the closest relative available and none to any of the others.

Suppose each mouthful of cake was equally valuable, translated into offspring flesh in simple pro-rata fashion. Then clearly an individual should prefer that his whole cake should be translated into closely related flesh than distantly related flesh. Of course this simple pro-rata assumption would almost certainly be false in real cases. However, quite elaborate assumptions about diminishing returns would have to be made before we could sensibly predict that the cake should be divided in exact proportion to coefficients of relatedness. Therefore, although my statement quoted above could under special circumstances be true, as a generalization it is properly regarded as fallacious. Of course I didn't really *mean* it anyway!

Apology

If the foregoing pages seem destructive or negative in tone, the very opposite was my intention. The art of explaining difficult material consists, in part, of anticipating the reader's difficulties and forestalling them. Systematically exposing common misunderstandings can therefore be a positively constructive exercise. I believe I understand kin selection better for having met these twelve errors, for having, in many cases, fallen into the trap myself and struggled painfully out the other side.

III

FUTURE CONDITIONAL

ROBERT WINSTON, in his thoughtful book *The Story of God*, ponders the distinction between the figures of 'priest' and 'prophet' in the history of religion: the former the rulemaker, the setter of boundaries, the enforcer; the latter the visionary, the critic, the refuser of false comfort, the grit in the communal oyster. But for Richard's protests, the present collection might have been entitled *Reason's Prophet*, and this group of pieces is about the scientist as prophet in that latter sense – the sense of being prepared to tread the tightrope between informed imagination and unfounded speculation, to 'think the unthinkable' and thereby *make* it thinkable. How does the past relate to the present, and how does either relate to possible futures? For a scientist, these questions fire up the engines of imagination; in the scientific mind, they are subject to the brake of scepticism.

The first piece here, 'Net gain', is a reply to the annual 'Brockman Question' posed by the founder of the online salon and intellectual hub *The Edge*, John Brockman. Drawing on a long-standing interest in computers, this essay not only celebrates the extraordinary – and extraordinarily rapid – burgeoning of the internet, but makes the breathtaking suggestion that, once communication between the elements of society becomes sufficiently fast, the boundary between 'individual' and 'society' may itself disintegrate, and individual human memory wither. On the way, it makes characteristically forceful observations about some cultural and political aspects of the internet's exponential growth, from the (poor) quality of much chat-room conversation to the (great) potential for freedom from oppressive authority it offers, via a fascinated glance at phenomena such as the taste for anonymity in communal interchange.

The second piece, 'Intelligent aliens', also has its origins in a Brockman initiative, this time a collection of essays on the 'intelligent design' movement. Here the focus shifts from possibilities of how human life might evolve further here on Earth to possibilities of contact with life forms further afield, in other parts of the universe. This particular foray out onto the tightrope represents the distinction between well-rooted speculation and declarative superstition – demonstrating, with a degree of irony, that the objective truth of science can send out imaginative probes just as daring as, and considerably better founded than, any form of supernaturalism. The next 'dart', 'Searching under the lamp-post', on the same subject but in lighter vein, takes a somewhat sceptical look at one approach to the search for extraterrestrial intelligence.

The final piece in this section both continues the thread of science-based speculation and sets out with unmistakable clarity a crucial distinction: that between the 'soul' as detachable inhabitant of an afterlife and the 'soul' as locus of the human spirit, its deep well of intellectual and emotional capacity; between the soul of established religion and wistful supernaturalism, and the soul as celebrated in the title of this collection, and in Richard's introduction to it. Under the provocative headline 'Fifty years on: killing the soul?', this essay delivers a ringing assertion of the aesthetic power and glory of the scientific vision alongside a brisk dismissal of any lingering Cartesian dualism. Science still has its mysteries, not least among them the nature of consciousness; but these are invitations to the scientists of the future, freed from supernaturalist constraints and set loose among the infinite possibilities of reality.

G.S.

Net gain*

How is the internet changing the way you think?

If, forty years ago, the Brockman Question had been: 'What do you anticipate will most radically change the way you think during the next forty years?', my mind would have flown instantly to a then recent article in *Scientific American* (September 1966) about 'Project MAC'. Nothing to do with the Apple Mac, which it long pre-dated, Project MAC was an MIT-based cooperative enterprise in pioneering computer science. It included the circle of artificial intelligence innovators surrounding Marvin Minsky but, oddly, that was not the part that captured my imagination. What really excited me, as a user of the large mainframe computers that were all you could get in those days, was something that nowadays would seem utterly commonplace: the then astonishing fact that up to thirty people simultaneously, from all around the MIT campus and even from their homes, could simultaneously log in to the same computer: simultaneously communicate with it and with each

* The literary agent John Brockman has the agreeable custom of mining his well-stocked address book every year around Christmas time and soliciting answers to an 'Annual Edge Question'. In 2011, the question was the topical one: 'How is the internet changing the way you think?' This was my contribution to the resulting volume.

other. *Mirabile dictu*, the co-authors of a paper could work on it simultaneously, drawing upon a shared database in the computer, even though they might be miles apart. In principle, they could be on opposite sides of the globe.

Today that sounds absurdly modest. It's hard to recapture how futuristic it was at the time. The post-Berners-Lee world, if we could have imagined it forty years ago, would have seemed shattering. Anybody with a cheap laptop computer, and an averagely fast wifi connection, can enjoy the illusion of bouncing dizzily around the world in full colour, from a beach webcam in Portugal to a chess match in Vladivostok, and Google Earth actually lets you fly the full length of the intervening landscape as if on a magic carpet. You can drop in for a chat at a virtual pub, in a virtual town whose geographical location is so irrelevant as to be literally non-existent (and the content of whose LOL-punctuated conversation, alas, is likely to be of a drivelling fatuity that insults the technology that mediates it).

'Pearls before swine' overestimates the average chat-room conversation, but it is the pearls of hardware and software that inspire me: the internet itself and the World Wide Web, succinctly defined by Wikipedia as 'a system of interlinked hypertext documents contained on the internet'. The Web is a work of genius, one of the highest achievements of the human species, whose most remarkable quality is that it was constructed not by one individual genius like Tim Berners-Lee or Steve Wozniak or Alan Kay, nor by a top-down company like Sony or IBM, but by an anarchistic confederation of largely anonymous units located (irrelevantly) all over the world. It is Project MAC writ large. Suprahumanly large. Moreover, there is not one massive central computer with lots of satellites, as in Project MAC, but a distributed network of computers of different sizes, speeds and manufacturers, a network that nobody, literally nobody, ever designed or put together, but which grew, haphazardly, organically, in a way that is not just biological but specifically *ecological*.

Of course there are negative aspects, but they are easily forgiven.

I've already referred to the lamentable content of many chat-room conversations without editorial control. The tendency to flaming rudeness is fostered by the convention – whose sociological provenance we might discuss one day – of anonymity. Insults and obscenities, to which you would not dream of signing your real name, flow gleefully from the keyboard when you are masquerading online as 'TinkyWinky' or 'FlubPoodle' or 'ArchWeasel'. And then there is the perennial problem of sorting out true information from false. Fast search engines tempt us to see the entire Web as a gigantic encyclopedia, while forgetting that traditional encyclopedias were rigorously edited and their entries authored by chosen experts. Having said that, I am repeatedly astounded by how good Wikipedia can be. I calibrate Wikipedia by looking up the few things I really do know about (and may indeed have written the entry for in traditional encyclopedias), say 'evolution' or 'natural selection'. I am so impressed by these calibratory forays that I go, with some confidence, to other entries where I lack first-hand knowledge (which was why I felt able to quote Wikipedia's definition of the Web, above). No doubt mistakes creep in, or are even maliciously inserted,* but the half-life of a mistake, before the natural correction mechanism kills it, is encouragingly short. John Brockman warns me that, while Wikipedia is indeed excellent on scientific matters, this is not always so 'in other areas such as politics and popular culture where . . . edit wars continually break out'. Nevertheless, the fact that the Wiki concept works, even if only in some areas such as science, flies so flagrantly in the face of all my prior pessimism that

---

* Insertions are sometimes more vain and self-serving than malicious. While performing my 'calibrating' read (see above) of the entry on natural selection, I noticed that the limited bibliography contained a book which I had read and knew was scarcely at all relevant to the subject. I went in and deleted it. Within half an hour it was back, I'm guessing reinserted by the author. I deleted it again. It returned again and I gave up, beaten. It isn't there, by the way, in the much longer and more thorough current entry.

I am tempted to see it as a metaphor for all that justifies optimism about the World Wide Web.

Optimistic we may be, but there is a lot of rubbish on the Web, more than in printed books, perhaps because they cost more to produce (and, alas, there's plenty of rubbish there too*). But the speed and ubiquity of the internet actually help us to be on our critical guard. If a report on one site sounds implausible (or too plausible to be true) you can quickly check it on several more. Urban legends and other viral memes are helpfully catalogued on various sites. When we receive one of those panicky warnings (often attributed to Microsoft or Symantec) about a dangerous computer virus, we preferably do *not* immediately spam it to our entire address book but instead Google a key phrase from the warning itself. It usually turns out to be, say, 'Hoax Number 76', its history and geography meticulously tracked.

Perhaps the main downside of the internet is that surfing can be addictive and a prodigious timewaster, encouraging a habit of butterflying from topic to topic, rather than attending to one thing at a time. But I want to leave negativity and naysaying and end with some speculative – perhaps more positive – observations. The unplanned worldwide unification that the Web is achieving (a science-fiction enthusiast might discern the embryonic stirrings of a new life form) mirrors the evolution of the nervous system in multicellular animals. A certain school of psychologists might see it as mirroring the development of each individual's personality, as a fusion among split and distributed beginnings in infancy. I am reminded of an insight that comes from Fred Hoyle's science-fiction novel *The Black Cloud*. The cloud is a superhuman interstellar traveller, whose 'nervous system' consists of units that communicate with each other by radio – orders of magnitude faster than our puttering nerve impulses. But in what sense is the cloud to be seen as a single individual rather than a society? The answer is that

---

* Especially now that computers make vanity publishing, with no editorial control, so cheap and easy.

interconnectedness that is sufficiently fast blurs the distinction. A human society would effectively become one individual if we could read each other's thoughts through direct, high-speed, brain-to-brain radio transmission. Something like that may eventually meld the various units that constitute the internet.

This futuristic speculation recalls the beginning of my essay. What if we look forty years into the future? Moore's Law will probably continue for at least part of that time, enough to wreak some astonishing magic (as it would seem to our puny imaginations if we could be granted a sneak preview today). Retrieval from the communal exosomatic memory will become dramatically faster, and we shall rely less on the memory in our skulls. At present we still need biological brains to provide the cross-referencing and association, but more sophisticated software and faster hardware will increasingly usurp even that function.

The high-resolution colour rendering of virtual reality will improve to the point where the distinction from the real world becomes unnervingly hard to notice. Large-scale communal games such as Second Life will become disconcertingly addictive to many who understand little of what goes on in the engine room. And let's not be snobbish about that. For many people around the world, 'first life' reality has few charms and, even for those more fortunate, active participation in a virtual world can be more intellectually stimulating than the life of a couch potato slumped in idle thrall to *Big Brother*. To intellectuals, Second Life and its souped-up successors will become laboratories of sociology, experimental psychology and *their* successor disciplines yet to be invented and named. Whole economies, ecologies and perhaps personalities will exist nowhere other than in virtual space.

Finally, there may be political implications. Apartheid South Africa tried to suppress opposition by banning television, and eventually had to give up. It will be more difficult to ban the internet. Theocratic or otherwise malign regimes may find it increasingly hard to bamboozle their citizens with their evil nonsense. Whether, on balance, the internet benefits the oppressed more than the oppressor is controversial, and at present may vary from region to region. We can at least hope that the faster, more ubiquitous and above all cheaper internet of the future may hasten the long-awaited downfall of ayatollahs, mullahs, popes, televangelists, and all who wield power through the control (whether cynical or sincere) of gullible minds. Perhaps Tim Berners-Lee will one day earn the Nobel Peace Prize.

AFTERWORD

Reading this again at the end of 2016 I find its generally optimistic tone a little jarring. There is alarmingly convincing evidence that the year's momentous US presidential election (it remains to be seen quite how momentous it will prove to be, not just for America but the world) was swayed by a systematically orchestrated campaign of fake news defaming one of the candidates. If further investigation proves this to be true, one would hope legislation, or at least self-policing by organizations such as Facebook and Twitter, will follow. At present these social media exult in their freedom of contribution as well as freedom of access. There's a minimum of editorial control, limited to censorship of gross obscenity and violent threats: no fact-checking such as that on which reputable newspapers like the *New York Times* pride themselves. There are already signs that reforms may be on the way. Unfortunately they'll be too late for the 2016 election.

Intelligent aliens*

Aᴍᴏɴɢ ᴛʜᴇ ᴍᴀɴʏ dishonesties of the well-financed intelligent-design cabal is the pretence that the designer is not the God of Abraham but an intelligence unspecified, who could equally well be an extraterrestrial alien.[†] Presumably the motive is to circumvent the First Amendment's prohibition on the establishment of religion, especially following Judge William Overton's 1982 decision in *McLean v. Arkansas Board of Education*, in which he struck down the state legislature's attempt to ensure 'balanced treatment' in the schools for 'creation science'.

The religious affiliation of these people is not in doubt, and their ingroup communications do not bother to hide their agenda. Jonathan Wells, one of the Discovery Institute's leading propagandists and the author of *Icons of Evolution*, is a lifelong member of the Unification Church (the Moonies). He wrote the following testimony in a Moonie in-house journal, under the heading 'Darwinism: why I went for a second PhD' (note that 'Father' is the Moonies' name for Reverend Moon himself):

* This is my contribution to another book edited by John Brockman, this time in 2006, entitled *Intelligent Thought: science versus the Intelligent Design movement*.

† This dishonesty frequently goes unnoticed. The intelligent design 'theorist' (theorist is too flattering a word) talks as though it's a minor detail whether the designer is God or an extraterrestrial alien. In fact the difference is huge, as this essay will show.

Father's words, my studies, and my prayers convinced me that I should devote my life to destroying Darwinism, just as many of my fellow Unificationists had already devoted their lives to destroying Marxism. When Father chose me (along with about a dozen other seminary graduates) to enter a PhD program in 1978, I welcomed the opportunity to prepare myself for battle.

This quotation alone casts doubt on any claim Wells might have had to be taken seriously as a disinterested seeker after truth – which would seem a fairly minimal qualification for a PhD in science. He publicly admits to undertaking a scientific research degree not in order to discover something about the world but for the specific purpose of 'destroying' a scientific idea that his religious leader opposed. Phillip Johnson, the born-again Christian law professor generally regarded as the leader of the gang, openly admits that his motive for opposing evolution is its 'naturalism' (as opposed to supernaturalism).

The claim that the intelligent designer might be an alien from outer space may be disingenuous, but this doesn't stop it from serving as the basis for an interesting and revealing discussion. Such a constructive discussion, *within* science, is what I shall undertake in this essay.

The problem of recognizing an alien intelligence arises, in its starkest form, in that branch of science known as SETI, the Search for Extra-Terrestrial Intelligence. SETI deserves to be taken seriously. Its practitioners are not to be confused with those who complain of having been abducted in flying saucers for sexual purposes. For all sorts of reasons including the reach of our listening devices and the speed of light, it is extremely unlikely that our first apprehension of an alien intelligence will be a corporeal visitation. SETI scientists do not anticipate meeting extraterrestrial visitors in the flesh but in the form of radio transmissions whose intelligent origin should, it is hoped, be evident from their patterning.

A strong case can be made for the probable existence of intelligent life elsewhere in the universe. It gains support from the principle of mediocrity, that salutary lesson from Copernicus, Hubble and others. Earth was once thought to be the only place in existence, surrounded by crystalline spheres bedecked with tiny stars. Later, when the size of the Milky Way galaxy was understood, it too was thought to be the only place, the locus of all that is. Then Edwin Hubble came along as a latter-day Copernicus to downgrade even our galaxy to mediocrity: it is only one among a hundred billion galaxies in the universe. Today, cosmologists look at our universe and seriously speculate that it may be one of many universes in the 'multiverse'.

Similarly, the history of our species was once thought to have been roughly coterminous with the history of everything. Now, to borrow Mark Twain's crushing analogy, our history's proportionate duration has shrunk to the thickness of the paint on top of the Eiffel Tower. If we apply the principle of mediocrity to life on this planet, doesn't it warn us that we would be foolhardy and vain to think that the Earth might be the only site of life in a universe of a hundred billion galaxies?

It is a powerful argument, and I find myself persuaded by it. On the other hand, the principle of mediocrity is emasculated by another powerful principle, known as the anthropic principle: the fact that we are in a position to observe the world's conditions determines that those conditions had to be favourable to our existence. The name comes from the British mathematician Brandon Carter, although he later preferred – with good reason – the 'self-selection principle'. I want to borrow Carter's principle for a discussion of the origin of life, the chemical event that forged the first self-replicating molecule and hence triggered natural selection of DNA and ultimately all of life. Suppose the origin of life really was a stupendously improbable event. Suppose the accident of primeval-soup chemistry which engendered the first self-replicating molecule was so prodigiously lucky that the odds against it were as low as one in a billion per billion planet years. Such fantastically low odds would mean that no chemist could entertain the smallest hope of repeating the event in a laboratory. The

National Science Foundation would laugh in the face of a research proposal whose admitted chance of success was as low as one in a hundred per year, let alone one in a billion per billion years. Yet so great is the number of planets in the universe that even these minuscule odds yield an expectation that the universe contains a billion planets bearing life. And (here comes the anthropic principle) since we manifestly live here, Earth necessarily has to be one of the billion.

Even if the odds against life arising on a planet are as low as one in a billion billion (which puts it well beyond the range we would classify as possible*), the plausible calculation that there are at least a billion billion planets in the universe provides an entirely satisfying explanation for our existence. There will still plausibly be one life-bearing planet in the universe. And once we have granted that, the anthropic principle does the rest. Any being contemplating the calculation necessarily has to be on that one life-bearing planet, which therefore has to be Earth.

This application of the anthropic principle is astonishing but watertight. I have oversimplified it by assuming that once life has originated on a planet, Darwinian natural selection will lead to intelligent and reflective beings. To be more precise, I should have been talking about the combined probability of life's originating on a planet and leading, eventually, to the evolution of intelligent beings capable of anthropic reflection. It could be that the chemical origin of a self-replicating molecule (the necessary trigger for the origin of natural selection) was a relatively probable event but later steps in the evolution of intelligent life were highly improbable. Mark Ridley, in *Mendel's Demon* (confusingly rebranded in America as *The Cooperative Gene*), suggests that the really improbable step

---

* Although it is strictly synonymous, I would now prefer to say 'Which puts it well into the range that we would classify as impossible.' Better still, 'for all practical purposes impossible'. When we are dealing with such huge numbers, 'possible', 'impossible' and 'practical purposes' have to be understood in impractical ways.

in our kind of life was the origin of the eucaryotic cell.* It follows from Ridley's argument that huge numbers of planets are home to something like bacterial life but only a tiny fraction of planets will have made it past the next hurdle to a level equivalent to the eucaryotic cell – what Ridley calls complex life. Or one might take the view that both those hurdles were relatively easy and that the really difficult step for terrestrial life was the attainment of the human level of intelligence. In this view, we would expect the universe to be rich in planets housing complex life but perhaps with only one planet harbouring beings capable of noticing their own existence and therefore of invoking the anthropic principle. It doesn't matter how we distribute our odds among these three 'hurdles' (or indeed other hurdles, such as the origin of a nervous system). So long as the total odds against a planet's evolving a life form capable of anthropic reflection do not exceed the number of planets in the universe, we have an adequate and satisfying explanation for our existence.

Although this anthropic argument is entirely watertight, my strong intuitive feeling is that we do not need to invoke it. I suspect that the odds in favour of life's arising and subsequently evolving intelligence are sufficiently high that many billions of planets do indeed contain intelligent life forms, many of them so superior to

* Eucaryotic cells are what we consist of – and by 'we' I mean all of life except bacteria and archaea. They are characterized by possession of a membrane-bound nucleus containing DNA, and 'organelles' such as mitochondria which, we now know, originated as symbiotic bacteria, and which still reproduce autonomously within the cell with their own DNA. Ridley is presumably right to regard such symbiotic unions as very improbable, lucky events. Nevertheless, there were two of them at least: one when green bacteria joined the club and provided – as chloroplasts – the photosynthetic know-how that all plants still use; and again when the ancestors of mitochondria came into the fold. Lynn Margulis (who had a track record of being right as well as wrong) believed there were yet more such momentous unions.

ours that we might be tempted to worship them as gods. Fortunately or unfortunately, we very likely won't encounter them: even such apparently high estimates still leave intelligent life marooned on scattered islands, which might well be, on average, too far apart for their inhabitants ever to visit one another. Enrico Fermi's famous rhetorical question 'Where are they?' could receive the disappointing answer: 'They are all over the place but too widely spaced to meet.' Nevertheless, my belief, for what it's worth, is that the odds against intelligent life are nothing like as great as the anthropic calculation allows us to countenance. And therefore I think it is well worth putting quite a lot of money into SETI. A positive result would be an exhilarating biological finding, equalled in the history of biology perhaps only by Darwin's discovery of natural selection itself.

If SETI ever does pick up a signal, it will likely be from the high, or godlike, end of the spectrum of cosmic intelligences.* We shall have a huge amount to learn from the aliens, especially about physics, which will be the same for them as it is for us although they'll know much more about it. Biology will be very different, though – just how different will be a fascinating question. Communication will be all one-way. If Einstein is right about the limiting speed of light, dialogue will be impossible. We may learn from them, but we won't be able to tell them about us in return.

How, then, would we recognize intelligence in a pattern of radio waves picked up by a giant parabolic dish and known to originate from deep space and not be a hoax? A tentative candidate was the pattern first detected by Jocelyn Bell Burnell in 1967 and jokingly called by her the LGM (Little Green Men) signal. This rhythmic pulse, with a periodicity of just over one second, is now known to have been a pulsar; indeed, hers was the first discovery of a pulsar. A pulsar is a neutron star rotating on its own axis with a

---

* Life forms of our present level don't have the technology adequate to traverse immense distances. So the barrier will have to be penetrated by beings with far superior technology and science.

beam of radio waves sweeping round and round as if from a lighthouse. That a star could rotate with a 'day' measured on a scale of seconds is an extremely surprising fact – not the only surprising fact about neutron stars. But for present purposes the important point is that the periodicity of Bell Burnell's signal is not an indicator of intelligent origin but an unaided product of ordinary physics. Plenty of very simple physical phenomena, from dripping water to pendulums of all kinds, are capable of yielding rhythmic patterns.

What next might occur to a SETI researcher as diagnostic of intelligent life? Well, if we assume that the aliens actively want to signal their presence, we can ask what we would do if we were trying to transmit evidence of our intelligent presence. Certainly not emit a rhythmic pattern like Bell Burnell's LGM signal, but what else? Several people have suggested prime numbers as the simplest kind of signal that could originate only from an intelligent source. But how confident should we be that a pattern of pulses based on prime numbers could come only from a mathematically sophisticated civilization? Strictly speaking, you can't prove that there is no inanimate physical system capable of generating prime numbers. You can say only that no physicist has ever discovered a non-biological process capable of generating them. Strictly speaking, the same caution goes for any signal. However, there are certain kinds of signals – of which those based on prime numbers may be the simplest example – which would be so convincing as to leave alternatives looking absurd.

Disquietingly, biologists have proposed models that are capable of generating prime numbers but do not involve intelligence. Periodical cicadas emerge for breeding every seventeen years (in some varieties) or every thirteen years (in other varieties). Two theories to account for this odd periodicity depend on the fact that 13 and 17 are prime numbers. I'll describe just one of these theories. Its premiss is that plague-year breeding is an adaptation to foil predators by swamping them. But then predator species evolved their own periodic breeding pattern to cash in on cicada plagues (or bonanzas, as they would see it). In an evolutionary arms race, the cicadas 'replied' by lengthening the period between plague years. The predators lengthened theirs in response. (Remember that this shorthand language of 'reply' and 'response' implies no conscious decisions, only blind natural selection.) When in the course of the arms race the cicadas reached an interval, such as six years, which was divisible by some other number, the predators found it more profitable to drop their breeding interval, to, say, three years, thereby hitting the cicada bonanza with alternate peaks of their own breeding cycle. Only when the cicadas hit a prime number did this become impossible. The cicadas continued to lengthen their stride until they reached a number that was too large to allow the predators to synchronize directly, yet prime and therefore impossible to meet with some multiple of a smaller period.

Well, that may not seem a very plausible theory, but it doesn't need to be for my purpose. I simply need to show that it is possible to conceive of a mechanistic model that does not involve conscious mathematics yet still manages to generate prime numbers. The cicada example shows that while prime numbers may not be generatable by non-biological physics, they can be generated by non-intelligent biology. Even the implausible cicada story is a cautionary tale to warn us that at least it is not necessarily obvious that prime numbers are diagnostic of intelligence.

The difficulty of diagnosing intelligence in a radio signal is itself a cautionary tale that calls to mind the historical analogy of the argument from design. There was a time when everybody (with a very few, very distinguished exceptions, such as David Hume) thought it completely obvious that the complexity of life was unmistakably diagnostic of intelligent design.* What should give us pause is this: Darwin's nineteenth-century contemporaries could claim the right to be as surprised by his remarkable discovery as we

---

* To revert to an earlier footnote (see page 152), this may be why nobody before Darwin and Wallace, not even the greatest thinkers such as Aristotle or Newton, tumbled to natural selection.

should be surprised today if a physicist discovered an inanimate mechanism capable of generating prime numbers. Perhaps we should entertain the possibility that other principles, comparable to Darwin's, remain to be discovered – principles capable of mimicking an illusion of design as convincing as the illusion manufactured by natural selection.

I am not inclined to predict any such event. Natural selection itself, properly understood, is powerful enough to generate complexity and the illusion of design to an almost limitless extent. Bear in mind that elsewhere in the universe there could be variants of natural selection that, although based on essentially the same principle as Darwin discovered on this planet, might be almost unrecognizably different in detail. Bear in mind, too, that natural selection can midwife other forms of design. It doesn't stop with its direct productions, such as feathers, ears and brains. Once natural selection has produced brains (or some extraterrestrial equivalent of brains), those brains can go on to produce technology (extraterrestrial equivalents of technology), including computers (or extraterrestrial equivalents of them) which are, like brains, capable of designing things. The manifestations of deliberate engineering design – indirect rather than direct productions of natural selection – can burgeon into new reaches of complexity and elegance. The point here is that natural selection manifests itself in the form of design at two levels: there is, first, the *illusion* of design, which we see in a bird's wing or a human eye or brain; and, second, there is 'true' design, which is a product of evolved brains.*

And now to my central point. There really is a profound difference

---

* My friend the philosopher Daniel Dennett argues forcefully, for example in *From Bacteria to Bach and Back*, that we should drop the word 'illusion' and simply use 'design' for what natural selection does. I see his point, but to follow him would obscure mine. In his terms we could say natural selection designs, and among the entities that it designs are those, such as brains, which are themselves capable of designing. I'm not going to get worked up about the semantics.

between an intelligent designer who is the product of a long period of evolution, whether on this planet or a distant one, and an intelligent designer who just *happened*, without any evolutionary history. When a creationist says that an eye or a bacterial flagellum or a blood-clotting mechanism is so complex that it must have been designed, it makes all the difference in the world whether the 'designer' is thought to be an alien produced by gradual evolution on a distant planet or a supernatural god who didn't evolve. Gradual evolution is a genuine explanation, which really can theoretically yield an intelligence of sufficient complexity to design machines and other things too complex to have come about by any process other than design. Hypothetical 'designers' jumped up from nothing cannot explain anything, because they can't explain themselves.

There are some man-made machines that commonsense, if not strict logic, tells us could not have come about by any process other than intelligent design. A jet fighter, a moon rocket, a motorcar, a bicycle – these are surely intelligently designed. But the important point is that the entity that did the designing – the human brain – is not. There is overwhelming evidence that the human brain evolved through a graded series of almost imperceptibly improving intermediates, whose relics may be seen in the fossil record and whose analogues survive all around the animal kingdom. Moreover, Darwin and his twentieth- and twenty-first-century successors have provided us with a luminously plausible explanation for the mechanism that propels evolution up the graded slopes, the process I have dubbed 'Climbing Mount Improbable'. Natural selection is not some desperate last resort of a theory. It is an idea whose plausibility and power hit you between the eyes with stunning force, once you understand it in all its elegant simplicity. Well might T. H. Huxley cry out, 'How extremely stupid of me not to have thought of that!'

But we can go further. Not only does natural selection explain the bacterial flagellum, the eye, the feather, and brains capable of intelligent design. Not only can it explain every biological phenomenon ever described. It is the only plausible explanation for these things that has ever been proposed. Above all, the argument from improbability – the very argument that the advocates of intelligent design fondly imagine supports their case – turns around to kick that case over with devastating force and lethal effect.

The argument from improbability states, incontrovertibly, that some phenomenon in nature – something like a bacterial flagellum, say, or an eye – is too improbable to have simply happened. It has to be the product of some very special process that generates improbability. The mistake is to jump to the conclusion that 'design' is that very special process. In fact it is natural selection. The late Sir Fred Hoyle's jocular analogy of the Boeing 747 is useful, although it, too, turns out to make the opposite point to the one he intended. The spontaneous origin of the complexity of life, he said, is as improbable as a hurricane blowing through a scrapyard and spontaneously assembling a Boeing 747. Everybody agrees that airliners and living bodies are too improbable to be assembled by chance. A more precise characterization of the kind of improbability we are talking about is *specified improbability* (or specified complexity). The 'specified' is important, for reasons that I explained in *The Blind Watchmaker*. I began by pointing out that randomly hitting the number that opens the large combination lock on a bank vault is improbable in the same sense as hurling scrap metal around and assembling an airliner:

> Of all the millions of unique and, with hindsight, equally improbable, positions of the combination lock, only one opens the lock. Similarly, of all the millions of unique and, with hindsight, equally improbable arrangements of a heap of junk, only one (or very few) will fly. The uniqueness of the arrangement that flies, or that opens the safe, is nothing to do with hindsight. It is specified in advance. The lock-manufacturer fixed the combination, and he has told the bank manager. The ability to fly is a property of an airliner that we specify in advance.

Given that chance is ruled out for sufficient levels of improbability, we know of only two processes that can generate specified improbability. They are intelligent design and natural selection, and only the latter is capable of serving as an ultimate explanation. It generates specified improbability from a starting point of great simplicity. Intelligent design can't do that, because the designer must itself be an entity at an extremely high level of specified improbability. Whereas the specification of the Boeing 747 is that it must be able to fly, the specification of the 'intelligent designer' is that it must be able to design. And intelligent design cannot be the ultimate explanation for anything, for it begs the question of its own origin.

From the lowlands of primeval simplicity, natural selection gradually and steadily ramps its way up the gentle slopes of Mount Improbable until, after sufficient geological time, the end product of evolution is an object such as an eye or a heart – something of such an elevated level of specified improbability that no sane person could attribute it to random chance. The single most unfortunate misunderstanding of Darwinism is that it is a theory of chance; the misunderstanding presumably stems from the fact that mutation is random.* But natural selection is anything but random. To escape from chance is the primary accomplishment that any theory of life must aspire to. Obviously, if natural selection were a theory of random chance, it could not be right. Darwinian natural selection is the *non*-random survival of randomly varying coded instructions for building bodies.

Some engineers even use explicitly Darwinian methods in order to optimize systems. They escalate performance from poor beginnings up a ramp of improvement to something approaching an optimum. Something like this process may be true of all engineers, even if they don't think of it as explicitly Darwinian. The engineer's wastepaper basket holds the 'mutant' designs he discarded before

---

* Actually that may be too charitable an explanation of the misunderstanding. It could stem from imaginations so impoverished as to think chance is by definition the default alternative to conscious design.

putting them to the test. Some designs don't even make it onto paper but are discarded in the engineer's head. I have no need to pursue the question of whether Darwinian natural selection is a good or helpful model for what goes on in the brain of a creative engineer or artist; constructive creative work – by engineers or artists, or indeed anybody – may or may not plausibly represent a form of Darwinism. The fundamental point remains that all specified complexity must ultimately rise from simplicity by some kind of escalatory process.

If we ever discover evidence that some aspect of life on Earth is so complex that it must have been intelligently designed, scientists will face with equanimity – and doubtless some excitement – the possibility that it was designed by an extraterrestrial intelligence. The molecular biologist Francis Crick, together with his colleague Leslie Orgel, made such a suggestion (I suspect, tongue in cheek) in proposing the theory of directed panspermia. According to Orgel and Crick's idea, extraterrestrial designers deliberately seeded Earth with bacterial life.* But the important point is that the designers were themselves the end product of some extraterrestrial version of Darwinian natural selection. The supernatural explanation fails to explain because it ducks the responsibility to explain itself.

Creationists who disguise themselves as 'intelligent-design theorists' have only one argument, and it goes like this:

1. The eye (the articulation of the mammalian jaw, the bacterial flagellum, the elbow joint of the lesser spotted weasel frog – which you have never heard of and don't

---

* I was once asked, in the course of a documentary which I didn't realize at the time was creationist propaganda, whether I could conceive of any way in which life on Earth could have been intelligently designed. I said that the only way (although I didn't believe it) would be design by an extraterrestrial intelligence, which would itself ultimately have to be the product of gradual evolution. I've never heard the end of it: 'Richard Dawkins believes in little green men.'

have time to look up before you seem to a lay audience
to have lost the argument) is irreducibly complex.

2. Therefore it cannot have evolved by gradual degrees.
3. Therefore it must have been designed.

No supporting evidence is ever offered for step 1, the allegation of irreducible complexity. I have sometimes referred to it as the argument from personal incredulity. It is always offered as a negative argument: theory A is alleged to fail in some respect, so we have to default to theory B, without even asking whether theory B might be deficient in the very same respect.

One legitimate response of biologists to the argument from personal incredulity is to attack step 2: look carefully at the examples proposed and show that they either did, or could, easily evolve by gradual degrees. Darwin did this for the eye. Later paleontologists did it for the articulation of the mammalian jaw. Modern biochemists have done it for the bacterial flagellum.

But the message of this essay is that, strictly speaking, we needn't bother to dispute steps 1 and 2. Even if they were ever accepted, step 3 remains irretrievably invalid. If incontrovertible evidence of intelligent design were ever discovered in, say, the organization of the bacterial cell – if we found evidence as strong as a manufacturer's signature written in unmistakable characters of DNA – this could only be evidence of a designer that was itself the product of natural selection or of some other as yet unknown escalatory process. If such evidence were found, our minds should immediately start working along the lines of Crick's directed panspermia, not a supernatural designer. Whatever else irreducible complexity might demonstrate, the one thing it cannot appeal to in ultimate explanation is something else that is irreducibly complex. Either you accept the argument from improbability, in which case it disproves the existence of ultimate designers. Or you don't accept it, in which case any attempt to deploy it against evolution is inconsistent if not dishonest. You cannot have it both ways.

AFTERWORD

Many theologians pathetically try to have it both ways through a piece of brazen assertion. By fiat they assert that their creator god is not himself complex and improbable, but simple. We know he is simple because eminent theologians like Thomas Aquinas say he is simple! Was there ever evasion more transparently evasive than this? Not only must any creator worthy of the name have the computational power to conceive the quantum physics of fundamental particles, the relativistic physics of gravity, the nuclear physics of stars and the chemistry of life. He also, at least in the case of Aquinas' God, has enough bandwidth left over to listen to the prayers and forgive – or not, according to taste – the sins of sentient beings all over his created universe. Simple?

Searching under the lamp-post*

THE JOKE IS familiar. Man searches diligently under lamp-post at night. Explains to passer-by that he has lost his keys. 'Did you lose them under the lamp-post?' 'No.' 'Then why are you looking under the lamp-post?' 'Because there's no light anywhere else.'

The argument has a certain zany logic, and it seems to appeal to Paul Davies, a distinguished British physicist now at Arizona State University. Davies is interested (as am I) in whether our kind of life is unique in the universe. The DNA code, the machine code of life, is all but identical in every living creature that has ever been examined. It is highly unlikely that the same 64-triplet code would coincidentally evolve more than once independently, and this is the main evidence that we are all cousins, sharing a single common ancestor, which probably lived between three and four billion years ago. If life originated more than once on this planet, only one life form survives: our kind of life, typified by our DNA code.

If there is life on other planets, it will very likely have something equivalent to a genetic code, but it is highly unlikely to be the same as ours. If we discover life, say on Mars, the acid test of whether it originated independently will be its genetic code. If it has DNA and the same 64-triplet DNA code, we shall conclude that it is a cross-contamination, perhaps via a meteorite.

We know that meteorites do occasionally travel between Earth and Mars – and, by the way, here is my second example of searching

---

* This article first appeared on the website of the Richard Dawkins Foundation for Reason and Science on 26 December 2011.

under the lamp-post. A meteorite can land anywhere on Earth, but we are unlikely to find it lying on any surface other than permanent snow: anywhere else it would just look like a stone, and it would soon be covered by vegetation or dust storms or soil movements. This is why scientists hunting for meteorites travel to the Antarctic: not because they are more likely to be there than anywhere else, but because that is where you can clearly see them even when they landed a long time ago. Antarctica is where the lamp-post is. Any stone or small rock lying on top of the snow must have dropped there – and it is quite likely to be a meteorite. Some meteorites found in Antarctica have been shown to come from Mars. This astonishing conclusion follows from a careful matching up of the chemical composition of these rocks with samples taken by robot spacecraft sent to Mars. Some time in the distant past, a large meteorite hit Mars with catastrophic impact. Fragments of Martian rock exploded up into space and some of them eventually ended up here. This shows that matter does sometimes travel between the two planets, and this opens up the possibility of cross-contamination by (presumably bacterial) life. If Earth-life did contaminate Mars (or vice versa), we would recognize it by its DNA code: it would be the same as ours.

Conversely, if we found a life form with a different genetic code – not DNA, or DNA with a different code – we would call it truly alien. Paul Davies suggests that maybe we don't need to go even as far as Mars to find truly alien life. Space travel is expensive and difficult. Maybe we should be searching right here for alien life that started on Earth, independently of ours, and never left. Maybe we should be systematically examining the genetic code of every micro-organism we can lay our hands on. Every one so far examined has the same genetic code as we do. But we have never systematically searched to see if we can find a different genetic code. Earth is Paul Davies' lamp-post because it is much cheaper and easier to search among Earthly bacteria than to travel to Mars, let alone to other star systems where the best hope of alien life reposes. I wish Paul good luck in his search under this particular lamp-post, but I am very doubtful of success, partly for the reason Charles Darwin himself gave: any other life form would probably have long ago been eaten by our kind – probably bacteria, we can today add.

I was reminded of all this by a news story in the *Guardian*:[*] 'Scientists to scour 1m lunar images for signs of alien life.' Yet again, the story concerns our old friend Paul Davies, and he is yet again down on his hands and knees, under yet another lamp-post.

If technologically advanced aliens ever visited us, they would be much more likely to have done so in the past than in the present, simply because the past is so much bigger than the present – if we define the present as one lifetime, or even as the span of recorded history. Traces of alien visitations – wrecked spacecraft, rubbish, evidence of mining activity, maybe even an intentionally deposited signal as in *2001: A Space Odyssey* – would quickly (by the standards of geological time) be covered over on the actively heaving and vegetation-covered surface of Earth. But the moon is another matter. No plants, no wind, no tectonic movements: Neil Armstrong walked in the lunar dust forty-two years ago, and his footprints probably still look fresh. So, Paul Davies and his colleague Robert Wagner reason, it makes sense to examine every high-resolution photograph ever taken of the moon's surface, just in case traces are to be seen.[†] The probability is low, but the payoff could be very high, so it is worth doing.

I am very sceptical. I suspect that there is life elsewhere in the universe, but it is probably extremely rare and isolated on far-flung islands of life, like a celestial Polynesia. Visitations to one island by another are hugely more likely to be in the form of radio transmissions than visitations by corporeal beings. This is because radio waves travel at the speed of light, whereas solid bodies travel

[*] 23 December 2011.

[†] The aliens of Arthur Clarke's story planted their tell-tale 'tombstone' signal on the moon so that it could be discovered only by a civilization advanced enough to be worthy of it.

only at the speed of – well, solid bodies. Moreover, radio waves travel outwards in an ever-expanding sphere, whereas bodies travel in only one direction at a time. This is why SETI (the Search for Extra-Terrestrial Intelligence using radio telescopes) is worthwhile. SETI is not wildly expensive as big science goes, but Paul Davies' latest lamp-post is a lot cheaper and I again wish him luck.

Fifty years on: killing the soul?*

FIFTY YEARS ON, science will have killed the soul. What a terrible, soulless thing to say! But only if you misunderstand it (easily done, admittedly). There are two meanings, Soul-1 and Soul-2, superficially confusable but deeply different. The following definitions from the *Oxford English Dictionary* convey what I am calling Soul-1.

> The spiritual part of man regarded as surviving after death, and as susceptible of happiness or misery in a future state.
>
> The disembodied spirit of a deceased person regarded as a separate entity and as invested with some amount of form and personality.

Soul-1, the soul that science is going to destroy, is supernatural, disembodied, survives the death of the brain and is capable of happiness or misery even when the neurones are dust and the hormones dry. Science is going to kill it stone dead. Soul-2, however, will never be threatened by science. On the contrary, science is its twin and handmaiden. These definitions, also from the *OED*, convey various aspects of Soul-2:

> Intellectual or spiritual power. High development of the mental

---

* Crystal-ball gazing is a notoriously error-prone indulgence but, for what it's worth, this was my contribution to Mike Wallace's 2008 edited book *The Way We Will Be Fifty Years from Today*.

faculties. Also, in somewhat weakened sense, deep feeling, sensitivity.

The seat of the emotions, feelings, or sentiments; the emotional part of man's nature.

Einstein was a great exponent of Soul-2 in science, and Carl Sagan was a virtuoso. *Unweaving the Rainbow* is my own modest celebration. Or listen to the great Indian astrophysicist Subrahmanyan Chandrasekhar:

> This 'shuddering before the beautiful', this incredible fact that a discovery motivated by a search after the beautiful in mathematics should find its exact replica in Nature, persuades me to say that beauty is that to which the human mind responds at its deepest and most profound.*

That was Soul-2, the kind of soulfulness that science courts and loves, and from which it will never be parted. The rest of this article refers only to Soul-1. Soul-1 is rooted in the dualistic theory that there is something non-material about life, some non-physical vital principle. It's the theory according to which a body has to be animated by an anima, vitalized by a vital force, energized by some mysterious energy, spiritualized by a spirit, made conscious by a mystical thing or substance called consciousness. It is no accident that all those characterizations of Soul-1 are circular. Julian Huxley memorably satirized Henri Bergson's *élan vital* by suggesting that a railway engine works by *élan locomotif* (incidentally, it is a lamentable fact that Bergson is still the only scientist ever to win the Nobel Prize in Literature). Science has already battered and wasted Soul-1. Within fifty years it will extinguish it altogether.

Fifty years back, we were only beginning to come to terms with Watson and Crick's 1953 paper in *Nature*, and few had tumbled to its poleaxing significance. Theirs was seen as no more than a clever

---

* Quoted in Martin Rees, *Before the Beginning*, p. 103. I used the same quotation in the first essay in this collection – but it bears repetition.

feat of molecular crystallography, while their last sentence ('It has not escaped our notice that the specific pairing we have postulated immediately suggests a possible copying mechanism for the genetic material') was just amusingly laconic understatement. With hindsight we can see that to call it understatement was itself the mother of understatements.

Before Watson/Crick (one contemporary scientist said to Crick, when Crick introduced him to Watson, 'Watson? But I thought your name was Watson-Crick') it was still possible for a leading historian of science, Charles Singer, to write:

> Despite interpretations to the contrary, the theory of the gene is not a 'mechanist' theory. The gene is no more comprehensible as a chemical or physical entity than is the cell or, for that matter, the organism itself . . . If I ask for a living chromosome, that is, for the only effective kind of chromosome, no one can give it to me except in its living surroundings any more than he can give me a living arm or leg. The doctrine of the relativity of functions is as true for the gene as it is for any of the organs of the body. They exist and function only in relation to other organs. Thus the last of the biological theories leaves us where the first started, in the presence of a power called life or psyche which is not only of its own kind but unique in each and all of its exhibitions.

Watson and Crick drove a coach and horses through all that: blew it ignominiously out of the water. Biology is becoming a branch of informatics. The Watson/Crick gene is a one-dimensional string of linear data, differing from a computer file only in the trivial respect that its universal code is quaternary not binary. Genes are isolatable strings of digital data, they can be read out of living or dead bodies, they can be written on paper and stored in a library, ready to be used again at any time. It is already possible, though expensive, to write your entire genome in a book, and mine in a similar book. Fifty years hence, genomics will be so cheap that the library (electronic library, of course) will house the complete genomes of as many individuals of as many thousands of species as we want. This will give us the final, definitive family tree of all life. Judicious comparison, in the library, of the genomes of any pair of modern species will allow us a fair shot at reconstructing their extinct common ancestor, especially if we also throw into the computational mix the genomes of its modern ecological counterparts. Embryological science will be so advanced that we'll be able to clone a living, breathing representative of that ancestor. Or of Lucy the Australopithecine, perhaps? Maybe even a dinosaur. And by 2057 it will be child's play to take down from its shelf the book that bears your name, type your genome back into a DNA synthesizer, insert it into an enucleated egg, and clone you – your identical twin but fifty years younger. Will it be a resurrection of your conscious being, a reincarnation of your subjectivity? No. We already know the answer is no, because monozygotic twins don't share a single subjective identity. They may have uncannily similar intuitions, but they do not think they are each other.

Just as Darwin in the mid-nineteenth century destroyed the mystical 'design' argument, and just as Watson and Crick in the mid-twentieth century destroyed all mystical nonsense about genes, their successors of the mid-twenty-first century will destroy the mystical absurdity of souls being detached from bodies. It won't be easy. Subjective consciousness is undeniably mysterious. In *How the Mind Works* Steven Pinker elegantly sets out the problem of consciousness, and asks where it comes from and what's the explanation. Then he's frank enough to say, 'Beats the heck out of me.' That's honest, and I echo it. We don't know. We don't understand it. Yet. But I believe we will, some time before 2057. And if we do, it certainly won't be mystics or theologians who solve this greatest of all riddles but scientists – maybe a lone genius like Darwin, but more probably a combination of neuroscientists, computer scientists and science-savvy philosophers. Soul-1 will die a belated and unlamented death at the hand of science, which will in the process launch Soul-2 to undreamed-of heights.

IV

MIND CONTROL, MISCHIEF AND MUDDLE

FOR ANY READERS STILL wondering why Richard Dawkins 'makes such a fuss' about religion, the title of this section hints at some of the reasons; the seven pieces that follow offer a more definitive answer from the apocalyptic horseman's mouth.

The first, 'The "Alabama Insert"', is a great set-piece demolition of creationism and reassertion of evolution by natural selection, and of the indispensable importance of scientific method. Originally delivered as a spur-of-the-moment defence of beleaguered educators faced with an attempt by the governing authorities to inhibit the teaching of genuine science, it should give pause to anyone who doubts the political force of creationism in present-day America.

From cool forensic analysis to the distillation of fury. The next piece, 'The guided missiles of 9/11', begins with deceptive calm, proceeding through passages of apparently technical description, then rises through a rapid crescendo of increasingly bitter irony to the punchline: the lethal force of irrational belief in a personal afterlife. Darts come no more pointed than this.

'The theology of the tsunami' shifts tone again, this time from anger to exasperation. In December 2004 a huge tsunami, generated by a powerful earthquake under the Indian Ocean, destroyed thousands of lives and livelihoods in South-East Asia. This account of the incomprehension of many religious people in the face of such undeserved suffering, the responses offered by religious leaders, and an ensuing exchange of correspondence on the *Guardian* letters page encapsulates several key elements of Richard's objection to religion, not least its misdirection of money, time, emotion and effort. Pointing out that an agonized 'Why?' was simply the wrong question (or rather, that it had a perfectly good answer in the geological rather than the theological realm), and that a more constructive response would be to 'get up off our knees, stop cringing before bogeymen and virtual fathers, face reality, and help science to do something constructive about human suffering', predictably won few plaudits among those unaccustomed to so bracing a challenge.

Lectures and letters loomed large among the candidate pieces for this collection: no accident, I think, for both offer immediacy of communication, whether to one person or to many simultaneously. The published open letter to an individual, of course, economically does both. 'Merry Christmas, Prime Minister!' takes the form of a seasonal greeting to David Cameron, at the time head of Britain's governing coalition. Making the case for a genuinely secular state in which, while individuals are free to adopt their own faiths, government remains scrupulously neutral, it robustly defends attachment to cultural myths, deriding the 'rebranding' of Christmas as a 'Winter Holiday', while pointing out the enduringly divisive effects of faith-based education and the inappropriateness – indeed, wickedness – of 'faith-labelling' of children. If we teach *about* religion rather than teaching *a* religion; if we understand our attachment to myth as what it is; if we are honest about where we get our ethics from and where we don't; then we'll all have happier Christmases.

Richard Dawkins is sometimes criticized for not taking religion seriously enough, for resorting too readily to dismissal rather than entering into genuine engagement. Leaving aside the evident seriousness of his excoriating attacks on the physical, psychological and educational harm wrought by religion, it is for its eagerness to interrogate the phenomenon of religion soberly, extensively and reflectively that I wanted to include here a substantial part of his 2005 lecture on 'The science of religion'. Readers of *The God Delusion* in particular will recognize some of the themes, argument and illustrations presented, but I make no apology for these echoes;

they fully merit recapitulation as a demonstration *par excellence* of the scientific lens applied to the cultural phenomenon. Here we see a patient, careful teasing out of the 'why' of faith and practice, showing the power of Darwinian natural selection as an explanatory tool, even – perhaps especially; certainly fittingly – when applied to belief systems that deny its efficacy. And one sentence sings out to me from this piece to epitomize the scientific method as practised by Dawkins, the demanding rigour of his approach to investigation: 'I am much more wedded to the general idea that the question should be properly put than I am to any particular answer.'

From a carefully refined question to a brisk and definitive answer: the next piece (also originally a lecture) disposes of the contention that 'belief' in science is itself a form of religion by reasserting the foundations of evidence, honesty and verifiability on which scientific investigation is based. It then moves on to more positive ground with a powerful reassertion of the virtues of science, explaining what science has to offer the human spirit in its hunger for explanation, its capabilities for astonishing feats of investigation, discovery, imagination and expression. Indeed, it suggests, let's teach science to children *in* their religious education classes – offer them not parochial superstition but the genuinely humbling visions of reality's own magic.

The section ends with a similarly positive and imaginative proposal in 'Atheists for Jesus': that we find a way to take what is good in religion *out* of religion and integrate it into the compassionate ethics of a secular society. Why should we not use our evolved big brains, our tendency to learn from and copy admired role models, in attempting a 'positive perversion' of Darwinian adaptation to spread 'superniceness'? Could there be an 'unselfishness meme'?

G.S.

The 'Alabama Insert'

PROLOGUE

Creationists believe that the biblical account of the creation of the universe is literally true; that God brought into existence the Earth and all its life forms in just six days. According to creationists, this event took place less than ten thousand years ago (they base their calculation of the age of the universe on the number of generations listed in the Bible – all those 'begats' strung together).

Creationists have succeeded in persuading large swathes of the general public that their theory is at least as scientifically respectable as the Big Bang/evolution alternative. Recent Gallup polls indicate that about 45 per cent of US citizens currently believe that God created human beings 'pretty much in (their) present form at one time or another within the last 10,000 years'.

In November 1995, the Alabama State Board of Education ordered that a one-page insert, labelled 'A Message from the Alabama State Board of Education', be stuck in all biology textbooks used in the state public schools. This flysheet formed the basis for a document used in the same way a little later in the state of Oklahoma. The 'Alabama Insert' is not exactly sophisticated, but it contains ritual gesturing in the direction of the educated reader. Above all, it says nothing about the religion that undoubtedly underlies it, and it pretends to the virtues of reasonable, scientific scepticism.

When I was invited to speak in Alabama around that time, a copy of the document was thrust into my hand before my lecture. I had also been made aware of the State Governor's recent performance on television. He had impersonated a shambling ape in an undignified attempt to ridicule the idea of evolution. I had the feeling that the biologists and honest educators of the State of Alabama felt embarrassed, threatened by their own state government, and in need of support. When I asked what they had to lose – why they didn't just teach evolution anyway – some admitted that they were literally fearful for their jobs, not just because of State interference but because of irate gangs of parents. On an impulse I threw aside my prepared script and devoted my lecture to dissecting the 'Alabama Insert' line by line, putting its successive clauses up on an overhead projector, there being no time to prepare slides. It is in a spirit of support for the beleaguered educators of Alabama, Oklahoma and other states and jurisdictions that I reproduce here an edited transcript of my remarks. Lines from the 'Alabama Insert' are printed in bold, followed by my responses.

This textbook discusses evolution, a controversial theory some scientists present as a scientific explanation for the origin of living things, such as plants, animals and humans.

This is misleading and disingenuous. 'Some' scientists, and 'controversial' theory, suggest the existence of a substantial number of respectable scientists who do not accept the fact of evolution. In reality the proportion of qualified scientists who do not accept evolution is minuscule. A few are paraded as possessing PhDs, but their PhDs are seldom from decent universities or in relevant subjects. Electrical and marine engineering are, no doubt, perfectly respectable disciplines, but their practitioners are no more qualified to pronounce on my subject than I am qualified to speak on theirs.

It is true that qualified biologists do not speak with one voice about every detail of evolution. Arguments will be heard in any flourishing branch of science. Not all biologists agree about the relative importance of Darwinian natural selection in guiding evolution, as compared with other possible forces such as genetic drift or higher-level quasi-Darwinian forces such as 'species selection'. But every reputable biologist, without exception, would accept the following proposition. All animals, plants, fungi and bacteria living today are descended from a single common ancestor who lived more than three billion years ago.* We are all cousins. This is not 'controversial' and it is not only 'some' scientists who believe it, except in the most narrowly pedantic meaning of the words. It is as near being a demonstrated fact as the theory that the alternation between night and day is caused by the rotation of the Earth. This leads into the next statement.

No one was present when life first appeared on Earth. Therefore, any statement about life's origins should be considered as theory, not fact.

The words 'theory' and 'fact' are here being used in a calculatedly misleading manner. Philosophers of science use the word 'theory' for pieces of knowledge that anybody else would call a fact, as well as for ideas that are little more than a hunch. It is a theory that 'mad cow disease' can be caught by humans as Creutzfeldt–Jakob Disease, a theory that might possibly be wrong; people are busy searching for further evidence, one way or the other. Various historical theories have been proposed as to the instigator of the Piltdown Man hoax, and we may never know the answer for certain. This is the common meaning of theory. But it is also technically a theory that the Earth is round, not flat. Just a theory that is overwhelmingly supported by the evidence.

---

* The only exception might be a respectable scientist such as Paul Davies (see page 208) who acknowledges the faint possibility that life may have arisen more than once, and that survivors, recognizable by their different genetic code, may still be among us. This conceivable exception in no way changes my statement. Purists could amend it to 'All known animals, plants . . .'.

The fact that nobody was present to witness the origins of life on Earth, or to witness the subsequent pageant of evolution, does not, in itself, bear decisively upon whether it should be considered fact. A murder may be unwitnessed, yet the circumstantial evidence of clues left behind, including fingerprints, footprints and DNA samples, can settle the culprit beyond all reasonable doubt. In science, lots of undoubted facts have never been witnessed directly, but they are more certain than many alleged direct observations. Nobody has lived long enough to see the continents move, but the theory of plate tectonics is overwhelmingly established, supported by a mass of evidence so large as to be beyond even unreasonable doubt. On the other hand, hundreds of eye-witnesses claim to have observed the sun miraculously changing direction at Fatima, at the behest of the Virgin Mary. Such eye-witness evidence cannot demonstrate that the sun really reversed itself, if only because the sun can be seen from much of the world at any one time and no eye-witness outside Fatima reported the event.*

According to the school of philosophy that is being implicitly invoked here, no 'fact' is ever more than a theory that has failed to be falsified after a massive battering of opportunities to falsify. If it makes you happy, I concede that evolution is only a theory, but it is a theory that is about as likely to be falsified as the theory that the Earth orbits the sun or the theory that Australia exists.

The word 'evolution' may refer to many types of change. Evolution describes changes that occur within a species (white moths, for example, may 'evolve' into gray moths). This process is micro-evolution, which can be observed and described as fact. Evolution may also refer to the change of one living thing to another, such as

---

* More to the point, if the sun had behaved in the way described by seventy thousand eye-witnesses in Fatima, our planet, if not the entire solar system, would have been destroyed. Eye-witness testimony is not all it's cracked up to be – a fact, incidentally, that trial juries need to understand better.

reptiles into birds. This process, called macro-evolution, has never been observed and should be considered a theory.

Predictably, the much-hyped distinction between micro-evolution and macro-evolution is becoming a hot favourite with creationists. It is easy to see why they have latched onto it, but actually it is an overrated distinction. This point is admittedly controversial, but many of us believe, in any case, that macro-evolution is nothing more than micro-evolution stretched out over a very long timespan. Let me clarify the matter.

Sexual reproduction sees to it that the genes of a population are a shuffled mix – the 'gene pool'. The range of individual bodies that we see at any one time is the outward and visible manifestation of the current gene pool. As the millennia go by, the gene pool may gradually change. Some genes become gradually more frequent in the pool, others less frequent. And the range of animals that we see changes accordingly. Perhaps the average specimen becomes taller, or shaggier, or darker in colour. They don't all become taller, there is still a good range of heights, but the distribution is shifted in the taller (or shorter) direction as the frequency profile of the gene pool changes.

That is micro-evolution, and a good deal is known about the underlying causes. Genes may change in frequency as the result of various chance processes. Or they can change in frequency in a more driven way, as the result of natural selection. Natural selection is the only known force that can produce improvement and the illusion of design. But, insofar as evolutionary change is not change for the better, there are plenty of other forces that might drive micro-evolution. For the moment, I shall talk of natural selection.

Individual animals with certain qualities – say shagginess in an encroaching ice age – are slightly more likely as a result to survive and reproduce. Therefore the genes that made them shaggy are slightly more likely to increase in frequency in the gene pool. This is why animals and plants become good at surviving and reproducing. Of course, what it takes to survive and reproduce varies among different species and in differing environments. The mole gene pool becomes loaded with mutually compatible genes that flourish in small furry crawling bodies digging for worms underground. The albatross gene pool becomes loaded with a different set of mutually compatible genes that flourish in large feathered bodies skimming the waves of the great southern oceans.

That is micro-evolution and our creationist friends are admitting that they have become reconciled to it. Instead, they are pinning their hopes on macro-evolution which, they have been encouraged to believe, is something completely different. It may be something completely different, but I doubt it. The great American paleontologist George Gaylord Simpson believed that macro-evolution is just micro-evolution writ large, writ slow and writ gradual over a sufficiently large number of thousands of generations. I agree with him, and am increasingly impressed by the speed with which gradualistic selection can accumulate to forge dramatic change. See, for instance, Jonathan Weiner's account, in *The Beak of the Finch*, of the research by Peter and Rosemary Grant on rapid evolution in 'Darwin's Finches' of the Galápagos Islands.

What is the alternative to the Simpson view? Some modern American paleontologists make much of an alleged 'decoupling' between micro-evolution – the slow, gradual change in gene frequencies within a gene pool – and macro-evolution, which they see as a relatively abrupt springing into existence of new species. Except insofar as I shall return to the matter when dealing with other sentences from the Alabama Insert, it is not necessary to air these controversies. They are matters of detail, which do not bear upon the fact of evolution itself. For the moment I'll simply record the justified annoyance with which the leading exponents of decoupled macro-evolution and 'punctuation' view the creationists attempts to hijack their brainchild. Stephen Gould, for example, says:

> Since we proposed punctuated equilibria to explain trends, it is infuriating to be quoted again and again by creationists – whether

through design or stupidity I do not know – as admitting that the fossil record includes no transitional forms . . . Duane Gish writes, 'According to Goldschmidt, and now apparently according to Gould, a reptile laid an egg from which the first bird, feathers and all, was produced.' Any evolutionist who believed such nonsense would rightly be laughed off the intellectual stage; yet the only theory that could ever envision such a scenario for the origin of birds is creationism – with God acting in the egg . . . I am both angry at and amused by the creationists; but mostly I am deeply sad.

I agree, but I incline towards being angry more than sad or amused.

Evolution also refers to the unproven belief that random, undirected forces produced a world of living things.

It is remarkable how common is this travesty of the Darwinian theory. Any idiot can see that, if Darwinism were really a random force, it could not possibly generate the elegantly adapted complexity of life. It is no surprise, therefore, that propagandists with their own reasons for wishing to discredit the theory would put it about that Darwinism is nothing more than random 'chance'. It is then easy to ridicule the theory by calculating how many shakes of a dice would be equivalent to the spontaneous springing into being of, say, an eye. Since natural selection is very much *not* a chance process, shaking dice is completely irrelevant.

But the sentence from the Insert uses the word 'undirected' as a synonym for random, and this needs more thoughtful handling. Natural selection is certainly not a random process, but is it 'directed'? No, if directed means guided by deliberate, conscious, intelligent intention. No, if directed means aimed at some future goal or target. But yes, if directed means leading to adaptive improvement; yes, if directed means giving rise to a superficially convincing illusion of brilliant design. For this, natural selection assuredly does. Darwin's achievement was not to denigrate the elegance of the illusion of design but to explain that it is an illusion.

There are many unanswered questions about the origin of life which are not mentioned in your textbook, including:

- **Why did the major groups of animals suddenly appear in the fossil record (known as the 'Cambrian Explosion')?**

We are extremely lucky to have any fossils at all. After an animal dies, many conditions have to be met if it is to become a fossil, and one or other of these conditions usually is not met. I personally would consider it an honour to be fossilized, but I don't have much hope of it.

It is particularly difficult for animals without hard skeletons to be fossilized.* Therefore, we wouldn't ordinarily expect to see the soft ancestors of animals that eventually evolved hard skeletons. We'd *expect* fossils to appear suddenly, when hard skeletons arose.

There are rare, exceptional circumstances in which the soft parts of animals are preserved. One of the outstanding examples is the fossil bed known as the Burgess Shale, in Canada. The Burgess Shale, together with a similar area in China, is the best fossil bed we have from the Cambrian era. What must have happened is that the ancestors of these animals evolved by gradual degrees before the Cambrian era, but didn't fossilize.

As I said, we are lucky to have any fossils at all. But in any case, it is misleading to think that fossils are the most important evidence for evolution. Even if we had no fossils whatsoever, the evidence for evolution from other sources would still be overwhelmingly strong.

---

* The turbellarian flatworms are a large, beautiful and flourishing class of animals. There are about as many species of Turbellaria as there are mammals, yet not a single turbellarian fossil has ever been found. Creationists presumably believe the Turbellaria have lived on Earth for the same length of time as all other animals, give or take a day or two during October 4004 BC. So if a massive class of animals failed to leave a single fossil, surely the vertebrates can be forgiven for a few 'gaps' in their fossil record.

- **Why have no new major groups of living things appeared in the fossil record for a long time?**

Major groups do not, and *should* not (according to the Darwinian theory) 'appear' in the fossil record. On the contrary, they should gradually evolve from earlier ancestors. Anyone would think that new phyla are supposed to spring spontaneously into existence.[*] Some forms of creationism might have them springing spontaneously into existence, but not Darwinism. The major divisions of the animal kingdom, the phyla, started out, mostly in the Precambrian, as different *species*.[†] Then they gradually diverged and diverged. A little bit later they became distinct genera. Then distinct families, then distinct orders, and so on. You wouldn't expect new phyla to 'arise' in recent times because by the time we see them they haven't had time to diverge far enough away from

---

[*] This was exactly the misapprehension formed by the distinguished (and far from stupid) theoretical biologist Stuart Kauffman, who imagined that 'species which founded taxa appear to have built up the higher taxa from the top down. That is, exemplars of major phyla were present first, followed by progressive filling in at class, order, and lower taxonomic levels.' This profound misunderstanding was nurtured by the excesses of 'poetic science' beloved of Stephen Jay Gould – specifically, Gould's book *Wonderful Life* – against which I warned in the Afterword to the essay on 'Universal Darwinism' in section II of this collection.

[†] It is surprising but true. Even more surprising, before they became separate species, the ancestors of any two of today's phyla were once offspring of the same mother. Take a human and a snail, for example. If you trace our ancestors back sufficiently far, and trace the snail's ancestors back sufficiently far, you will eventually converge on a single individual, the common ancestor of them both. One child of this parent was destined to give rise to us (and all the vertebrates plus starfish and some worms). Another child of this parent was destined to give rise to snails (and insects, most worms, lobsters, octopuses etc.).

their ancestors to be recognized as distinct phyla. Come back in five hundred million years and birds, for instance, may have evolved so far from the other vertebrates that they will be classified as belonging in their own phylum.

As an analogy, think of an old oak tree with major boughs bearing small twigs. Every major bough began life as a small twig. If somebody said to you: 'Isn't it odd, no new major boughs have sprung from this tree for a long time? All we've had in recent years is new small twigs,' you'd think they were pretty stupid, wouldn't you? Well, yes, stupid is the word.

- **Why do major new groups of plants and animals have no transitional forms in the fossil record?**

It is astonishing how frequently this is stated in creationist literature. I don't know where it came from, because it simply isn't true. It seems to be sheer wishful thinking. In fact, just about every fossil found is potentially an intermediate between something and something else. There are gaps too, for the reasons I have stated. But what there is not is a single example of a fossil in the *wrong* place. The great British biologist J. B. S. Haldane was once challenged, by a zealous proponent of Karl Popper's philosophy that science proceeds by proposing *falsifiable* hypotheses, to name a single discovery which would falsify the theory of evolution. 'Fossil rabbits in the Precambrian,' Haldane growled. No such misplaced fossil has ever been authentically found.

All the fossils that we have are in the right order. Creationists know this and see it as an awkward fact that needs explaining. The best explanation they can come up with is truly bizarre. It all came about because of Noah's Flood. The animals understandably tried to save their skins by heading for the hills. As the waters rose, the cleverest animals held out longest and reached higher up the slopes before drowning. That's why we find fossils of 'higher' animals above fossils of 'lower' animals. Well, *ad hoc* explanations don't come more piteously desperate than that.*

Part of the creationist error about gaps in the fossil record may come from a gleeful misunderstanding of the theory of punctuated equilibrium, propounded by Eldredge and Gould. Eldredge and Gould were talking about a jerkiness in the fossil record, which stems from the fact that, on their view of evolution, most evolutionary change takes place relatively rapidly, during what they call speciation events. Between speciation events there are long periods of stasis during which no evolutionary change occurs. It is a ludicrous confusion to muddle this up – as creationists wilfully do – with major gaps in the fossil record such as that which preceded the so-called Cambrian Explosion. I have already quoted Dr Gould's justified annoyance at being persistently misquoted by creationists.

Finally, there is a purely semantic point about classification. I can explain it best with an analogy. Children turn gradually and continuously into adults but, for legal purposes, the age of majority is taken to be a particular birthday, often the eighteenth. It would therefore be possible to say: 'There are fifty-five million people in Britain but not a single one of them is intermediate between non-voter and voter. There are no intermediates: an embarrassing gap in the developmental progression.' Just as, for legal purposes, a juvenile changes into a voter as midnight strikes on their eighteenth birthday, so zoologists always insist on classifying a specimen as in one species or another. If a specimen is intermediate in actual form (as many are, in accordance with Darwinian expectations),

---

* Even a member of the Alabama State Legislature might be capable of understanding that an explanation of that kind can, in any case, only ever be *statistical*, not absolute. The 'head for the hills' theory might explain why there is a statistical preponderance of advanced animals in higher strata. But the trend could only be statistical. In actual fact, there is not a single exception to the rule, not one single solitary example of a mammal fossil, say, in a stratum too low in the fossil record.

zoologists' legalistic conventions still force them to jump one way or the other. Therefore the creationist's claim that there are no intermediates has to be true *by definition* at the species level, but it has no implications about the real world – only implications about zoologists' naming conventions.

The proper way to look for intermediates is to forget the *naming* of fossils and look, instead, at their actual shape and size. When you do that, you find that the fossil record abounds in beautifully gradual transitions, although there are some gaps too – some very large and accepted, by *everybody*, as due to animals simply failing to fossilize. To look no further than our own ancestry, the transition from *Australopithecus* to *Homo habilis* to *Homo erectus* to 'archaic *Homo sapiens*' to 'modern *Homo sapiens*' is so smoothly gradual that fossil experts are continually squabbling about where to classify – how to name – particular fossils. Now look at the following, from a book of anti-evolution propaganda: 'The finds have been referred to as either *Australopithecus* and hence are apes, or *Homo* and hence are human. Despite more than a century of energetic excavation and intense debate the glass case reserved for mankind's hypothetical ancestor remains empty. The missing link is still missing.' One is left wondering what a fossil has to do to qualify as an intermediate. What could it *conceivably* do? In fact the statement quoted is saying nothing whatever about the real world.

- **How did you and all living things come to possess such a complete and complex set of 'instructions' for building a living body?**

The set of instructions is our DNA. We got it from our parents, and they got it from their parents, and so on back to a tiny, remote ancestor, simpler than a bacterium, which lived about four thousand million years ago in the sea.

Since all organisms inherit all their genes from their ancestors, rather than from their ancestors' unsuccessful contemporaries, all organisms tend to possess successful genes. They have what it takes to become ancestors – and that means to survive and reproduce.

This is why organisms tend to inherit genes with a propensity to build a well-designed machine: a body which actively works as if it is striving to become an ancestor. That is why birds are so good at flying, fish so good at swimming, monkeys so good at climbing, viruses so good at spreading. That is why we love life and love sex and love children. It is because we all, without a single exception, inherit all our genes from an unbroken line of successful ancestors. *The world becomes full of organisms that have what it takes to become ancestors.*

There's a lot more to it than that. Evolutionary arms races, such as the race run in evolutionary time between predators and their prey, or parasites and their hosts, have led to escalating perfection and complexity. As predators become better equipped to catch prey, so prey have become better equipped to evade capture. This is why antelopes and cheetahs both run so fast. It is why they are so good at detecting each other's presence. Many details of a cheetah's or an antelope's body can be understood if you realize that each is the end product of a long evolutionary arms race against the other.

Study hard and keep an open mind. Someday, you may contribute to the theories of how living things appeared on Earth.

Well, at last I've found something I can agree with.

The guided missiles of 9/11*

A CONVENTIONAL GUIDED MISSILE corrects its trajectory as it flies, homing in, say, on the heat of a jet plane's exhaust. A great improvement on a simple ballistic shell, it still cannot distinguish particular targets. It could not zero in on a designated New York skyscraper if launched from as far away as Boston.

That is precisely what a modern 'smart missile' can do. Computer miniaturization has advanced to the point where one of today's smart missiles could be programmed with an image of the Manhattan skyline together with instructions to home in on the north tower of the World Trade Center. Smart missiles of this sophistication are possessed by the United States, as we learned in the Gulf War, but they are economically beyond ordinary terrorists and scientifically beyond theocratic governments. Might there be a cheaper and easier alternative?

In the Second World War, before electronics became cheap and miniature, the psychologist B. F. Skinner did some research on pigeon-guided missiles. The pigeon was to sit in a tiny cockpit, having previously been trained to peck keys in such a way as to keep a designated target in the centre of a screen. In the missile, the target would be for real.

The principle worked, although it was never put into practice

---

* Reactions to the religious crime now universally known as 9/11 were varied and passionate. I wrote several, of which this was the first, published in the *Guardian* just four days after the event.

by the US authorities. Even factoring in the costs of training them, pigeons are cheaper and lighter than computers of comparable effectiveness. Their feats in Skinner's boxes suggest that a pigeon, after a regimen of training with colour slides, really could guide a missile to a distinctive landmark at the southern end of Manhattan island. The pigeon has no idea that it is guiding a missile. It just keeps on pecking at those two tall rectangles on the screen, from time to time a food reward drops out of the dispenser, and this goes on until . . . oblivion.

Pigeons may be cheap and disposable as on-board guidance systems, but there's no escaping the cost of the missile itself. And no such missile large enough to do much damage could penetrate US airspace without being intercepted. What is needed is a missile that is not recognized for what it is until too late. Something like a large civilian airliner, carrying the innocuous markings of a well-known carrier and a great deal of fuel. That's the easy part. But how do you smuggle on board the necessary guidance system? You can hardly expect the pilots to surrender the left-hand seat to a pigeon or a computer.

How about using humans as on-board guidance systems, instead of pigeons? Humans are at least as numerous as pigeons, their brains are not significantly costlier than pigeon brains, and for many tasks they are actually superior. Humans have a proven track record in taking over planes by the use of threats, which work because the legitimate pilots value their own lives and those of their passengers.

The natural assumption that the hijacker ultimately values his own life too, and will act rationally to preserve it, leads air crews and ground staff to make calculated decisions that would not work with guidance modules lacking a sense of self-preservation. If your plane is being hijacked by an armed man who, though prepared to take risks, presumably wants to go on living, there is room for bargaining. A rational pilot complies with the hijacker's wishes, gets the plane down on the ground, has hot food sent in for the passengers and leaves the negotiations to people trained to negotiate.

The problem with the human guidance system is precisely this. Unlike the pigeon version, it knows that a successful mission culminates in its own destruction. Could we develop a biological guidance system with the compliance and dispensability of a pigeon but with a man's resourcefulness and ability to infiltrate plausibly? What we need, in a nutshell, is a human who doesn't mind being blown up. He'd make the perfect on-board guidance system. But suicide enthusiasts are hard to find. Even terminal cancer patients might lose their nerve when the crash was actually looming.

Could we get some otherwise normal humans and somehow persuade them that they are not going to die as a consequence of flying a plane smack into a skyscraper? If only! Nobody is that stupid, but how about this – it's a long shot, but it just might work. Given that they are certainly going to die, couldn't we sucker them into believing that they are going to come to life again afterwards? Don't be daft! No, listen, it might work. Offer them a fast track to a Great Oasis in the Sky, cooled by everlasting fountains. Harps and wings wouldn't appeal to the sort of young men we need, so tell them there's a special martyr's reward of seventy-two virgin brides, guaranteed eager and exclusive.

Would they fall for it? Yes, testosterone-sodden young men too unattractive to get a woman in this world might be desperate enough to go for seventy-two private virgins in the next.

It's a tall story, but worth a try. You'd have to get them young, though. Feed them a complete and self-consistent background mythology to make the big lie sound plausible when it comes. Give them a holy book and make them learn it by heart. Do you know, I really think it might work. As luck would have it, we have just the thing to hand: a ready-made system of mind control which has been honed over centuries, handed down through generations. Millions of people have been brought up in it. It is called religion and, for reasons which one day we may understand, most people fall for it (nowhere more so than America itself, though the irony passes unnoticed). Now all we need is to round up a few of these faith-heads and give them flying lessons.

Facetious? Trivializing an unspeakable evil? That is the exact opposite of my intention, which is deadly serious and prompted by deep grief and fierce anger. I am trying to call attention to the elephant in the room that everybody is too polite – or too devout – to notice: religion, and specifically the devaluing effect that religion has on human life. I don't mean devaluing the life of others (though it can do that too), but devaluing one's own life. Religion teaches the dangerous nonsense that death is not the end.

If death is final, a rational agent can be expected to value his life highly and be reluctant to risk it. This makes the world a safer place, just as a plane is safer if its hijacker wants to survive. At the other extreme, if a significant number of people convince themselves, or are convinced by their priests, that a martyr's death is equivalent to pressing the hyperspace button and zooming through a wormhole to another universe, it can make the world a very dangerous place. Especially if they also believe that that other universe is a paradisical escape from the tribulations of the real world. Top it off with sincerely believed, if ludicrous and degrading to women, sexual promises, and is it any wonder that naive and frustrated young men are clamouring to be selected for suicide missions?

There is no doubt that the afterlife-obsessed suicidal brain really is a weapon of immense power and danger. It is comparable to a smart missile, and its guidance system is in many respects superior to the most sophisticated electronic brain that money can buy. Yet to a cynical government, organization or priesthood, it is very, very cheap.

Our leaders have described the recent atrocity with the customary cliché: mindless cowardice. 'Mindless' may be a suitable word for the vandalizing of a telephone box. It is not helpful for understanding what hit New York on 11 September 2001. Those people were not mindless and they were certainly not cowards. On the contrary, they had sufficiently effective minds braced with an insane courage, and it would pay us mightily to understand where that courage came from.

It came from religion. Religion is also, of course, the underlying source of the divisiveness in the Middle East which motivated the use of this deadly weapon in the first place. But that is another story and not my concern here. My concern here is with the weapon itself. To fill a world with religion, or religions of the Abrahamic kind, is like littering the streets with loaded guns. Do not be surprised if they are used.

The theology of the tsunami*

I HAVE NEVER FOUND the problem of evil very persuasive as an argument against the existence of deities. There seems to be no obvious reason to presume that your God will be good. The question for me is why one thinks any God, good or evil or indifferent, exists at all. Most of the Greek pantheon sported very human vices, and the 'jealous God' of the Old Testament is surely one of the nastiest, most truly evil characters in all fiction.[†] Tsunamis would be just up his street, and the more misery and mayhem, the better. I have always thought the 'problem of evil' was a relatively trivial difficulty for theists, compared to the argument from improbability, which is a genuinely powerful, indeed, knockdown argument against the very existence of all forms of unevolved creative intelligence.

Nevertheless, my experience is that godly people who show no evidence of even beginning to understand the argument from improbability are reduced to quivering embarrassment, if not

* I let the habit slip, rather to my regret, but for a few years I was a regular columnist in *Free Inquiry*, one of two excellent journals published by the Center for Inquiry (the CFI has this year merged with my own Foundation, I am happy to say). This is one of my columns, published in 2005 soon after the terrible tsunami of Boxing Day 2004, which wrought widespread devastation on coastal areas around the Indian Ocean.

† See Dan Barker's *God: the most unpleasant character in all fiction* for ample justification of this judgement.

outright loss of faith, when confronted with a natural disaster or a major pestilence. Earthquakes, in particular, have traditionally shaken people's faith in a deity, and December's tsunami provoked a lot of agonized soul-searching on the question: 'How can religious people explain something like this?' The most prominent apparent quaverer was the archbishop of Canterbury, the head of the Anglican communion. It turned out that he had been traduced by the *Daily Telegraph*, a notoriously irresponsible and mischievous newspaper and one of several published in London that devoted many column inches to this knotty theological conundrum. The archbishop had not in fact said that the tsunami shook his own faith, only that he could sympathize with those who did have doubts.

The most famous precedent, several commentators reminded us, is the Lisbon earthquake of 1755, which deeply disturbed Kant and provoked Voltaire's mockery of Leibniz, and his philosophical optimism, in *Candide*. The *Guardian* published a flurry of letters to the editor, headed by one from the bishop of Lincoln, who asked God to preserve us from religious people who try to 'explain' the tsunami. Other letter writers attempted just that. One clergyman conceded that there was no intellectual answer, just hints of an explanation that 'will only be found in a life lived by faith, prayer, contemplation and Christian action'. Another clergyman cited the Book of Job, and thought he had found the beginnings of an explanation for suffering in Paul's idea that the whole universe was experiencing something akin to the pains of a woman in childbirth: 'The argument for the existence of God from design would be fatally flawed if the universe were seen as complete already. Religious believers see the totality of experience as part of a greater narrative moving towards an as yet unimaginable goal.'

Is this the kind of thing theologians are paid to do? At least he didn't sink to the level of a professor of theology in my university, who once suggested, during a televised discussion with me and my colleague Peter Atkins, among others, that the Holocaust was God's way of giving the Jews the opportunity to be brave and noble. That remark prompted Dr Atkins to growl, 'May you rot in hell!'

My own initial response to the correspondence on the tsunami was published on 30 December:

> The Bishop of Lincoln (Letters, December 29) asks to be preserved from religious people who try to explain the tsunami disaster. As well he might. Religious explanations for such tragedies range from loopy (it's payback for original sin) through vicious (disasters are sent to try our faith) to violent (after the Lisbon earthquake of 1755, heretics were hanged for provoking God's wrath). But I'd rather be preserved from religious people who give up on trying to explain, yet remain religious.
>
> In the same batch of letters, Dan Rickman says 'science provides an explanation of the mechanism of the tsunami but it cannot say why this occurred any more than religion can'. There, in one sentence, we have the religious mind displayed before us in all its absurdity. In what sense of the word 'why' does plate tectonics not provide the answer?
>
> Not only does science know why the tsunami happened, it can give precious hours of warning. If a small fraction of the tax breaks handed out to churches, mosques and synagogues had been diverted into an early warning system, tens of thousands of people, now dead, would have been moved to safety.
>
> Let's get up off our knees, stop cringing before bogeymen and virtual fathers, face reality, and help science to do something constructive about human suffering.

Letters to the editor necessarily have to be brief, and I failed to insure myself against the obvious charge of callousness. Among the onslaught that flooded the letters page the next day, one woman wondered what comfort science could offer to a parent whose child had been swept out to sea. Three letters were from doctors, who could justly claim more experience of human suffering than I could match. One of them deployed a bizarrely literal-minded interpretation of Darwinism: 'If I were an atheist, I can't imagine why I should bother to help anyone whose genes might compete with mine.' Another lashed petulantly out at science 'cloning sheep or cats'. The third attacked me personally, describing me as his personal bogeyman: 'the atheist version of a door-stepping Jehovah's Witness. An ayatollah without a deity – God help us.'

I don't usually come back for a second go, but I was anxious to dispel wanton misunderstanding, so I sent in another letter which was published the following day:

> It is true that science cannot offer the consolations that your correspondents attribute to prayer, and I am sorry if I seemed a callous ayatollah or a doorstepping bogeyman (Letters, December 31). It is psychologically possible to derive comfort from sincere belief in a nonexistent illusion, but – silly me – I thought believers might be disillusioned with an omnipotent being who had just drowned 125,000 innocent people (or an omniscient one who failed to warn them). Of course, if you can derive comfort from such a monster, I would not wish to deprive you.
>
> My naive guess was that believers might be feeling more inclined to curse their God than pray to him, and maybe there's some dark comfort in that. But I was trying, however insensitively, to offer a gentler and more constructive alternative. You don't have to be a believer. Maybe there's nobody there to curse. Maybe we are on our own, in a world where plate tectonics and other natural forces occasionally cause appalling catastrophes. Science cannot (yet) prevent earthquakes, but science could have provided just enough warning of the Boxing Day tsunami to save most of the victims and spare the bereaved. Even worse, lowland floodings of the future are threatened by global warming which is preventable by human action, guided by science. And if the comforts afforded by outstretched human arms, warm human words and heartbroken human generosity seem puny against the agony, they at least have the advantage of existing in the real world.

One of the most popular religious responses to natural disasters is 'Why me?' It underlay several of the replies to the first of my letters to the *Guardian*. The correct answer, 'Unfortunately you happened to be in the wrong place at the wrong time,' is admittedly not much comfort. The world is divided into those who can see that the capacity to comfort has no bearing on the truth of a cosmic claim and those who cannot. When I, as a professional educator, meet one of the latter, I come close to despair.

AFTERWORD

If apparently undeserved natural disasters pose a challenge to the religious, apparently undeserved good fortune may be said to pose an equal and opposite challenge to the non-religious: whom are we to thank? And why, indeed, do we *want* to give thanks, just as we want to blame someone or something for our misfortunes? In a lecture I gave at the Global Atheist Convention in Melbourne in 2010, I suggested a Darwinian explanation for these impulses of gratitude and resentment based on the evolution of a sense of 'fairness'.*

> When a hurricane destroys our house but spares the house of a really vile criminal, we are overcome by a feeling of unfairness. When a twister roars across the plain and suddenly veers sideways in the nick of time, just when it was about to hit our town, we feel an overwhelming sense of gratitude. We feel an urge to thank somebody, or some thing. Perhaps we don't thank the hurricane itself (which we have the sense to realise isn't listening), but we might thank 'Providence' or 'Fate' or something that we might call 'God' or 'the gods' or 'Allah', or whatever name our society gives to the target of such gratitude. And if the twister doesn't veer out of the way, and does destroy our house and kill our family, we may then cry out to the same god or gods, and perhaps say something like, 'What did I do to deserve this?' Or we might say, 'This must be my punishment for sinning, this is the payback for my sins.'
>
> Disasters can also, strangely enough, be the subject of

---

* For more on how this sense of 'natural justice' might have arisen, see the first essay in this collection, 'The values of science and the science of values', especially pages 55–9.

gratitude. Hundreds of thousands of people may die, in an earthquake or a tsunami, yet if one child is lost, presumed dead, and then discovered clinging to a piece of driftwood, the parents will feel an overwhelming urge to *thank* somebody, or some thing, that their child was restored to them after being presumed dead.

The urge to feel 'grateful' in a vacuum, when there's nobody there to thank, is very strong. Animals sometimes perform complicated patterns of behaviour in a vacuum – they are even called 'vacuum activities'. The most spectacular example I know is from a German film I once saw of a beaver. This was a captive beaver, but I must first remind you of something that wild beavers do. They build dams, mostly of logs or branches, which they cut to size using their very sharp gnawing teeth and push into the growing dam. You might wonder why they build dams, by the way. The reason is that the dam makes a lake or pond, which helps them to find their food without being eaten. Beavers probably don't understand why they do it. They just do it without thinking, because they have a mechanism in the brain that goes off like clockwork. They are like little dam-building robots. The clockwork behaviour patterns that form the components of dam-building routines are quite complicated and very different from the movements that any other animal does – because no other animal builds dams.

Now, the beaver in the German film was a captive beaver, which had never built a real dam in its life. It was filmed in a bare room, with a bare cement floor: no river to dam, and no wood to dam it with. But, amazingly, this poor lonely beaver went through all the motions of building a dam in a vacuum. It would pick up phantom pieces of wood in its jaws, and carry them to its phantom dam, shoving them in, tamping them down, generally behaving as though it 'thought' there was a real dam there, and real wood to tamp into it.

I think this beaver felt an overwhelming urge to build a dam, because that is what it would do in nature. And it went ahead and 'built' a phantom dam in a vacuum. I think what the beaver felt must have been a bit like what a man feels when he lusts over a picture of a naked woman – it perhaps gives him an erection – yet he knows perfectly well that it is only printing ink on paper. It is vacuum lust. What I am now suggesting is that we also feel vacuum gratitude. It is the gratitude we feel when we are overwhelmed with the urge to 'thank' something or someone, even though there is nobody to thank. It is gratitude in a vacuum, just like the beaver's dam-building in a vacuum. And the same goes for the way we feel when we say 'unfair' even though we know that there is nobody there to blame for the unfairness: we just feel hard done by, by the weather, or by an earthquake, or by 'fate'.

So, that is a possible evolutionary reason why we feel an urge to give thanks, even when we know there is nobody to thank. It is nothing to be ashamed of.

'Thank' shouldn't have to be a transitive verb. We don't have to thank God, or Allah, or the saints, or the stars. We can simply be thankful, and that's just fine.

Merry Christmas, Prime Minister!*

Dear Prime Minister,

Merry Christmas! I mean it. All that 'Happy Holiday Season' stuff, with 'holiday' cards and 'holiday' presents, is a tiresome import from the US, where it has long been fostered more by rival religions than by atheists. A cultural Anglican (whose family has been part of the Chipping Norton Set since 1727, as you'll see if you look around you in the parish church[†]), I recoil from secular carols such as 'White Christmas', 'Rudolph the Red-Nosed Reindeer' and the loathsome 'Jingle Bells', but I'm happy to sing real

---

* In November 2011 the *Guardian* invited a number of people to pose questions to the then Prime Minister, David Cameron, and Mr Cameron replied in a subsequent issue of the newspaper. I was one of those invited and my question was a serious and polite question about faith schools. Mr Cameron's rudely dismissive reply, in which he accused me of 'not getting it', stung me into writing an open reply in the Christmas 2011 issue of *New Statesman*, of which I was guest editor. My original title was 'Do you get it now, Prime Minister?' but I here retitle it in friendlier vein.

† Note for non-British readers: David Cameron was the Member of Parliament for West Oxfordshire, which includes my home town of Chipping Norton. He, and a number of other prominent members of the London political and journalistic class, have country homes in the area, and had beome known in the gossip columns as the 'Chipping Norton Set'. The church, as I unkindly implied he might have noticed if he were as pious as he pretends, is loaded to the architraves with Dawkins family memorials.

carols, and in the unlikely event that anyone wants me to read a lesson I'll gladly oblige – only from the King James version, of course.

Token objections to cribs and carols are not just silly, they distract vital attention from the real domination of our culture and politics that religion still gets away with, in (tax-free) spades. There's an important difference between traditions freely embraced by individuals, and traditions enforced by government edict. Imagine the outcry if your government were to require every family to celebrate Christmas in a religious way. You wouldn't dream of abusing your power like that. And yet your government, like its predecessors, does force religion on our society, in ways whose very familiarity disarms us. Setting aside the twenty-six bishops in the House of Lords, passing lightly over the smooth inside track on which the Charity Commission accelerates faith-based charities to tax-free status while others (quite rightly) have to jump through hoops, the most obvious and most dangerous way in which governments impose religion on our society is through faith schools.

We should teach *about* religion, if only because religion is such a salient force in world politics and such a potent driver of lethal conflict. We need more and better instruction in comparative religion (and I'm sure you'll agree with me that any education in English literature is sadly impoverished if the child can't take allusions from the King James Bible). But faith schools don't so much teach *about* religion as indoctrinate in the particular religion that runs the school. Unconscionably, they give children the message that they belong specifically to one particular faith, usually that of their parents, paving the way, at least in places such as Belfast and Glasgow, for a lifetime of discrimination and prejudice.

Psychologists tell us that, if you experimentally separate children in any arbitrary way – say, dress half of them in green T-shirts and half in orange – they will develop ingroup loyalty and outgroup prejudice. To continue the experiment, suppose that, when they grow up, greens only marry greens and oranges only marry oranges. Moreover, 'green children' only go to green schools and 'orange children' to orange schools. Carry on for three hundred years and what have you got? Northern Ireland, or worse. Religion may not be the only divisive power that can propel dangerous prejudices down through many generations (language and race are other candidates) but religion is the only one that receives active government support in Britain today in the form of schools.

So deeply ingrained is this divisive ethos in our social consciousness that journalists, and indeed most of us, breezily refer to 'Catholic children', 'Protestant children', 'Muslim children', 'Christian children', even where the children are too young to decide what they think about questions that divide the various faiths. We assume that children of Catholic parents (for instance) just *are* 'Catholic children', and so on. A phrase such as 'Muslim child' should grate like fingernails on a blackboard. The appropriate substitution is 'child of Muslim parents'.

I satirized the faith-labelling of children in the *Guardian* last month,* using an analogy that almost everybody gets as soon as he hears it – we wouldn't dream of labelling a child a 'Keynesian child' simply because her parents were Keynesian economists. Mr Cameron, you replied to that serious and sincere point with what could distinctly be heard on the audio version as a contemptuous snigger: 'Comparing John Maynard Keynes to Jesus Christ shows, in my view, why Richard Dawkins just doesn't really get it.' Do you get it now, Prime Minister? Obviously I was not comparing Keynes with Jesus. I could just as well have used 'monetarist child' or 'fascist child' or 'postmodernist child' or 'Europhile child'. Moreover, I wasn't talking specifically about Jesus, any more than Muhammad or the Buddha.

In fact, I think you got it all along. If you are like several government ministers (of all three parties) to whom I have spoken,

‖   *   26 November 2011.

you are not really a religious believer yourself. Several ministers and ex-ministers of education whom I have met, both Conservative and Labour, don't believe in God but, to quote the philosopher Daniel Dennett, they do 'believe in belief'. A depressingly large number of intelligent and educated people, despite having outgrown religious faith, still vaguely presume without thinking about it that religious faith is somehow 'good' for other people, good for society, good for public order, good for instilling morals, good for the common people even if we chaps don't need it. Condescending? Patronizing? Yes, but isn't that largely what lies behind successive governments' enthusiasm for faith schools?

Baroness Warsi, your Minister without Portfolio (and without election) has been at pains to inform us that this coalition government does indeed 'do God'.* But we who elected you mostly do not. It is possible that the recent census may register a slight majority of people ticking the 'Christian' box. However, the UK branch of the Richard Dawkins Foundation for Reason and Science commissioned an Ipsos MORI poll in the week following the

---

* Sayeeda Warsi, whose only known distinction was failing to win election to Parliament, was elevated to the peerage as the youngest member of the House of Lords by David Cameron, and made co-chair of the Conservative Party and a government minister. Rightly or wrongly, this was widely interpreted as three-way tokenism – she was the first female non-white Muslim member of the British Cabinet. My dig may have been unfair (though I doubt it), but in any case I feel the need of a footnote to explain it to non-British readers who might not otherwise have got it. Mr Cameron would certainly have got it, in the event (which again I doubt) that he had time to read my open letter. The phrase 'do God' is an allusion to the previous government of Tony Blair, whose chief spin doctor, Alastair Campbell, embarrassed by his boss's leanings towards the pious, interrupted a religious question during an interview by saying 'We don't do God.'

census. When published, this will enable us to see how many people who self-identified as Christian are *believers*.*

Meanwhile, the latest British Social Attitudes survey, just published, clearly demonstrates that religious affiliation, religious observance and religious attitudes to social issues have all continued their long-term decline and are now irrelevant to all but a minority of the population. When it comes to life choices, social attitudes, moral dilemmas and sense of identity, religion is on its deathbed, even for many of those who still nominally identify with a religion.

This is good news. It is good news because if we depended on religion for our values and our sense of cohesion we would be well and truly stuck. The very idea that we might get our morals from the Bible or the Quran will horrify any decent person today who takes the trouble to read those books – rather than cherry-pick the verses that happen to conform to our modern secular consensus. As for the patronizing assumption that people need the promise of heaven (or the obscene threat of torture in hell) in order to be moral, what a contemptibly immoral motive for being moral! What binds us together, what gives us our sense of empathy and compassion – our goodness – is something far more important, more fundamental and more powerful than religion: it is our common humanity, deriving from our pre-religious evolutionary heritage, then refined and improved, as Professor Steven Pinker

---

* It is now published, and I summarized the results in the tenth anniversary edition of *The God Delusion*. Briefly, the percentage of people who self-identified as Christian dropped dramatically between 2001 and 2011, and our survey showed that even those who still did so in 2011 were Christian only very nominally. For example, the dominant answer to the question what being a Christian meant to them was: 'I try to be a good person.' Yet, when asked whether they took religion into account when facing a moral choice, only 10 per cent said yes. Only 39 per cent of the self-identifying Christians were able to name which of the following four is the first book of the New Testament: Matthew, Genesis, Psalms, Acts.

argues in *The Better Angels of our Nature*, by centuries of secular enlightenment.*

A diverse and largely secular country such as Britain should not privilege the religious over the non-religious, or impose or underwrite religion in any aspect of public life. A government that does so is out of step with modern demographics and values. You seemed to understand that in your excellent, and unfairly criticized, speech on the dangers of 'multiculturalism' in February this year.†

Modern society requires and deserves a truly secular state, by which I mean *not* state atheism, but state *neutrality* in all matters pertaining to religion: the recognition that faith is personal and no business of the state. Individuals must always be free to 'do God' if they wish; but a government for the people certainly should not.

With my best wishes to you and your family for a happy Christmas,

Richard Dawkins

---

* I have tried to go a little further into this in the Afterword to the first piece in this collection (see page 64).

† I've since learned that Mr Cameron's speech was written with the advice of the admirable Maajid Nawaz of the Quilliam Foundation. Not surprising it was so good, therefore.

The science of religion*

IT IS WITH trepidation and humility that I come from the oldest university in the English-speaking world to what must surely be the greatest. My trepidation is not lessened by the title that, perhaps unwisely, I gave the organizers all those months ago. Anybody who publicly belittles religion, however gently, can expect hate mail of a uniquely unforgiving species. But the very fact that religion arouses such passions catches a scientist's attention.

As a Darwinian, the aspect of religion that catches *my* attention is its profligate wastefulness, its extravagant display of baroque uselessness. If a wild animal habitually spends time performing some useless activity, natural selection will favour rival individuals who devote the time, instead, to promoting their own survival or reproduction. Nature cannot afford time-wasting frivolity. Ruthless utilitarianism trumps, even if it doesn't always seem that way.

I am a Darwinian student of animal behaviour – an ethologist and follower of Niko Tinbergen. You won't be surprised, therefore, if I talk about animals (nonhuman animals, I should add, for there

---

* The Tanner Lectures on Human Values were founded in Cambridge in 1978, with the unusual remit of rotating around a number of different university venues. I have given Tanner Lectures in Edinburgh and Harvard. My two Harvard lectures, given in 2003, were a symmetrical pair called 'The science of religion' and 'The religion of science'. The first of them, in abbreviated form, is reproduced here.

is no sensible definition of an animal that excludes ourselves). The tail of a male bird of paradise, extravagant though it seems, would be penalized by females if it were less so. The same for the time and labour that a male bower bird puts into making his bower. Anting is the odd habit of birds, such as jays, of 'bathing' in an ant's nest and apparently inciting the ants to invade the feathers. Nobody knows for sure what the benefit of anting is: perhaps some kind of hygiene measure, cleansing the feathers of parasites. My point is that uncertainty as to detail doesn't – nor should it – stop Darwinians believing, with great confidence, that anting must be *for* something.

Such a confident stance is controversial – at Harvard, if nowhere else – and you may be aware of the wholly unwarranted slur that functional hypotheses are untestable 'Just So Stories'. This is such a ridiculous claim that the only reason it has come to be widely accepted is a certain style of bullying advocacy originating, I reluctantly have to say, at Harvard. All you have to do to test a functional hypothesis of a piece of behaviour is to engineer an experimental situation in which the behaviour doesn't happen, or in which its consequences are negated. Let me give a simple example of how to test a functional hypothesis.

Next time a housefly lands on your hand, don't immediately brush it off; watch what it does. You won't wait long before it brings its hands together as if in prayer, then wrings them in what seems like ritual fastidiousness. This is one of the ways in which a fly grooms itself. Another is to wipe a hind leg over the same side wing. They also rub middle and hind feet together, or middle and front. Flies spend so much time self-grooming, any Darwinian would immediately guess that it is vital for survival. The more so because – this is less paradoxical than it sounds – grooming is quite likely to lead directly to the fly's death. When a chameleon, for example, is around, grooming is very likely to be the last thing the fly does. Predatory eyes often lock onto movement. A motionless target goes unnoticed, even totally unseen. A flying target is difficult to hit. A grooming fly's shuttling limbs stimulate the predator's movement-detectors, but the fly as a whole is a sitting target. The fact that flies spend so much time grooming, in spite of its being so dangerous, argues for a very strong survival value. And this is a testable hypothesis.

An appropriate experimental design is the 'Yoked Control'. Put a matched pair of flies in a small arena and watch them. Every time Fly A starts to groom itself, scare both flies into flight. At the end of two hours of this regime, Fly A will have done no grooming at all. Fly B will have groomed itself a great deal. It will have been scared off the ground as many times as A, but at random with respect to its grooming. Now put A and B through a battery of comparison tests. Is A's flying performance impaired by dirty wings? Measure it and compare it with B's. Flies taste with their feet, and it is a reasonable hypothesis that 'foot washing' unclogs their sense organs. Compare the threshold sugar concentration that A and B can taste. Compare their tendency to disease. As a final test, compare the two flies' vulnerability to a chameleon.

Repeat the trial with lots of pairs of flies and do a statistical analysis comparing each A with its corresponding B. I would bet my shirt on the A flies being significantly impaired in at least one faculty vitally affecting survival. The reason for my confidence is purely the Darwinian conviction that natural selection would not have allowed them to spend so much time on an activity if it were not useful. This is not a 'Just So Story'; the reasoning is thoroughly scientific, and it is fully testable.*

Religious behaviour in bipedal apes occupies large quantities of time. It devours huge resources. A medieval cathedral would consume hundreds of man-centuries in the building. Sacred music and devotional paintings largely monopolized medieval and Renaissance talent. Thousands, perhaps millions, of people have died, often accepting torture first, for loyalty to one religion rather than a scarcely distinguishable alternative.

---

* My confidence definitely does not reside in any particular hypothesis such as that dirty wings impair flight. I am confident only that grooming must do something to improve the flies' genetic survival, simply because they spend so much time doing it.

Though the details differ across cultures, no culture is known that does not practise some version of the time-consuming, wealth-consuming, hostility-provoking, fecundity-forfeiting rituals of religion. All this presents a major puzzle to anyone who thinks in a Darwinian way. Isn't religion a challenge, an *a priori* affront to Darwinism, demanding similar explanation? Why do we pray and indulge in costly practices which, in many cases, more or less totally consume our lives?

Could religion be a recent phenomenon, sprung up since our genes underwent most of their natural selection? Its ubiquity argues against any simple version of this idea. Nevertheless there is a version of it that it will be my main purpose to advocate today. The propensity that was naturally selected in our ancestors was not religion *per se*. It had some other benefit, and it only incidentally manifests itself as religious behaviour. We'll understand religious behaviour only after we have renamed it. Once again, it is natural for an ethologist to use an example from non-human animals.

The 'dominance hierarchy' was first discovered as the 'pecking order' in hens. Each hen learns which individuals she can beat in a fight, and which beat her. In a well-established dominance hierarchy, little overt fighting is seen. Stable groupings of hens, who have time to sort themselves into a pecking order, lay more eggs than hens in coops whose membership is continually changed. This might suggest an 'advantage' to the phenomenon of the dominance hierarchy. But that's not good Darwinism, because the dominance hierarchy is a group-level phenomenon. Farmers may care about group productivity, but natural selection doesn't.

For a Darwinian, the question, 'What is the survival value of the dominance hierarchy?' is an illegitimate question. The proper question is: 'What is the individual survival value of deferring to stronger hens and punishing lack of deference from weaker ones?' Darwinian questions have to direct attention towards the level at which genetic variations might exist. Tendencies to aggression or deference in individual hens are a proper target because they either do or easily might vary genetically. Group phenomena like dominance hierarchies don't in themselves vary genetically, because groups don't have genes. Or at least, you'll have your work cut out arguing some peculiar sense in which a group phenomenon could be subject to genetic variation. You might contrive it via some version of what I have called the 'extended phenotype', but I am too sceptical to accompany you on that theoretical journey.

My point, of course, is that the phenomenon of religion may be like the dominance hierarchy. 'What is the survival value of religion?' may be the wrong question. The right question may have the form, 'What is the survival value of some as yet unspecified individual behaviour, or psychological characteristic, which manifests itself, under appropriate circumstances, as religion?' We have to rewrite the question before we can sensibly answer it.

I must first acknowledge that other Darwinians have gone straight for the un-rewritten question, and proposed direct Darwinian advantages of religion itself – as opposed to psychological predispositions that accidentally manifest themselves as religion. There is a little evidence that religious belief protects people from stress-related diseases. The evidence is not good, but it would not be surprising. A non-negligible part of what a doctor can provide for a patient is consolation and reassurance. My doctor doesn't literally practise the laying on of hands. But many's the time I have been instantly cured of some minor ailment by a reassuringly calm voice from an intelligent face surmounting a stethoscope. The placebo effect is well documented. Dummy pills, with no pharmacological activity at all, demonstrably improve health. That is why drug trials have to use placebos as controls. It's why homoeopathic remedies appear to work, even though they're so dilute that they have the same amount of the active ingredients as the placebo control – zero molecules.

Is religion a medical placebo, which prolongs life by reducing stress? Perhaps, although the theory is going to have to run the gauntlet of sceptics who point out the many circumstances in which religion increases stress rather than decreases it. In any case, I find the placebo theory too meagre to account for the massive and all-pervasive worldwide phenomenon of religion. I do not think we have religion because our ancestors reduced their stress levels and hence survived a little longer. I don't think that's a big enough theory for the job.

Other theories miss the point of Darwinian explanations altogether. I mean suggestions such as: 'Religion satisfies our curiosity about the universe and our place in it.' Or: 'Religion is consoling. People fear death and are drawn to religion because of the promise of surviving it.' There may be some psychological truth here, but it's not in itself a Darwinian explanation. A Darwinian version of the fear-of-death theory would have to be of the form, 'Belief in survival after death tends to postpone the moment when it is put to the test.' This could be true or it could be false – maybe it's another version of the stress and placebo theory – but I shall not pursue the matter. My only point is that this is the *kind* of way in which a Darwinian must rewrite the question. Psychological statements – that people find some belief agreeable or disagreeable – are proximate, not ultimate, explanations.

Darwinians make much of the distinction between proximate and ultimate. Proximate questions lead us into physiology and neuroanatomy. There is nothing wrong with proximate explanations. They are important, and they are scientific. But my preoccupation today is with Darwinian ultimate explanations. If neuroscientists, such as the Canadian Michael Persinger, find a 'god centre' in the brain, Darwinian scientists like me want to know why the god centre evolved. Why did those of our ancestors who had a genetic tendency to grow a god centre survive better than rivals who did not?

Some alleged ultimate explanations turn out to be – or in some cases avowedly are – group selection theories. Group selection is the controversial idea that Darwinian selection chooses among groups of individuals, in the same kind of way as it chooses among individuals within groups.

Here's a made-up example, to show what a group-selection theory of religion might look like. A tribe with a stirringly belligerent 'god of battles' wins wars against a tribe whose god urges peace and harmony, or a tribe with no god at all. Warriors who believe that a martyr's death will send them straight to paradise fight bravely and willingly give up their lives. So tribes with certain kinds of religion are more likely to survive in intertribal selection, steal the conquered tribe's cattle and seize their women as concubines. Such successful tribes spawn daughter tribes who go off and propagate more daughter tribes, all worshipping the same tribal god. Notice that this is different from saying that the *idea* of the warlike religion survives. Of course it will, but in this case the point is that the group of people who hold the idea survive.

There are formidable objections to group selection theories. A partisan in the controversy, I must beware of riding off on a hobby horse far from today's subject. There is also much confusion in the literature between true group selection, as in my hypothetical example of the god of battles, and something else that is *called* group selection but turns out to be either kin selection or reciprocal altruism. Or there may be a confusion of 'selection between groups' and 'selection between individuals in the particular circumstances furnished by group living'.

Those of us who object to group selection have always admitted that in principle it can happen. The problem is that, when it is pitted against individual-level selection – as when group selection is advanced as an explanation for individual self-sacrifice – individual-level selection is likely to be stronger. In our hypothetical tribe of martyrs, a single self-interested warrior, who leaves martyrdom to his colleagues, will end up on the winning side because of their gallantry. Unlike them, however, he ends up alive, outnumbered by women, and in a conspicuously better position to pass on his genes than his fallen comrades.

Group selection theories of individual self-sacrifice are always vulnerable to subversion from within. If it comes to a tussle between the two levels of selection, individual selection will tend to win because it has a faster turnover. Mathematical models arguably come up with special conditions under which group selection might work. Arguably, religions in human tribes set up just such

THE SCIENCE OF RELIGION

special conditions. This is an interesting line of theory to pursue, but I shall not do so here.

Instead, I shall return to the idea of rewriting the question. I previously cited the pecking order in hens, and the point is so central to my thesis that I hope you will forgive another animal example to ram it home. Moths fly into the candle flame, and it doesn't look like an accident. They go out of their way to make a burnt offering of themselves. We could label it 'self-immolation behaviour' and wonder how Darwinian natural selection could possibly favour it. My point, again, is that we need to rewrite the question before we can even attempt an intelligent answer. It isn't suicide. Apparent suicide emerges as an inadvertent side-effect.

Artificial light is a recent arrival on the night scene. Until recently, the only night lights were the moon and the stars. Because they are at optical infinity, their rays are parallel, which makes them ideal compasses. Insects are known to use celestial objects to steer accurately in a straight line.* They can use the same compass, with reversed sign, for returning home after a foray. The insect nervous system is adept at setting up a temporary rule of thumb such as 'Steer a course such that the light rays hit your eye at an angle of 30°.' Since insects have compound eyes, this will amount to favouring a particular ommatidium.†

But the compass relies critically on the celestial object being at

* For a wonderful example of this – how worker bees tell their fellows where to find food by reference to the sun – see the essay 'About time' in section VI of this collection (see page 331).

† Think of the compound eye as a hemispherical pincushion, densely covered with pins. Each 'pin' is actually a tube called an ommatidium, with a tiny photocell at its base. So an insect 'knows' the location of an object, such as the sun or a star, by which of its tubes is/are receiving light from that object. It is a very different kind of eye from our 'camera eye', whose image is upside down and left/right reversed. Insofar as a compound eye can be said to have an image at all, it is the right way up.

optical infinity. If it is not at infinity, the light rays are not parallel but diverge like the spokes of a wheel. A nervous system using a 30° rule of thumb to a candle, as though it were the moon, will steer its moth, in a neat logarithmic spiral, into the flame.

It is still, on average, a good rule of thumb. We don't notice the hundreds of moths who are silently and effectively steering by the moon or a bright star, or even the lights of a distant city. We see only moths hurling themselves at our lights, and we ask the wrong question: Why are all these moths committing suicide? Instead, we should ask why they have nervous systems that steer by maintaining an automatic fixed angle to light rays, a tactic that we notice only on the occasions when it goes wrong. When the question is rephrased, the mystery evaporates. It never was right to call it suicide.

Once again, apply the lesson to religious behaviour in humans. We observe large numbers of people – in many local areas it amounts to 100 per cent – who hold beliefs that flatly contradict demonstrable scientific facts as well as rival religions. They not only *hold* these beliefs but devote time and resources to costly activities that flow from holding them. They die for them, or kill for them. We marvel at all this, just as we marvelled at the 'self-immolation behaviour' of the moths. Baffled, we ask why. Yet again, the point I am making is that we may be asking the wrong question. The religious behaviour may be a misfiring, an unfortunate manifestation of an underlying psychological propensity that in other circumstances was once useful.

What might that psychological propensity have been? What is the equivalent of the parallel rays from the moon as a useful compass? I shall offer a suggestion, but I must stress that it is only an example of the kind of thing I am talking about. I am much more wedded to the general idea that the question should be properly put than I am to any particular answer.

My specific hypothesis is about children. More than any other species, we survive by the accumulated experience of previous generations. Theoretically, children might learn from experience not to swim in crocodile-infested waters. But, to say the least, there will be a selective advantage to child brains with a rule of thumb: Believe whatever your grown-ups tell you. Obey your parents, obey the tribal elders, especially when they adopt a solemn, minatory tone. Obey without question.

Natural selection builds child brains with a tendency to believe whatever their parents and tribal elders tell them. And this very quality automatically makes them vulnerable to infection by mind viruses. For excellent survival reasons, child brains need to trust parents, and trust elders whom their parents tell them to trust. An automatic consequence is that the truster has no way of distinguishing good advice from bad. The child cannot tell that 'If you swim in the river you'll be eaten by crocodiles' is good advice but 'If you don't sacrifice a goat at the time of the full moon, the crops will fail' is bad advice. They both sound equally trustworthy. They are both advice from a trusted source, both delivered with a solemn earnestness that commands respect and demands obedience.

The same goes for propositions about the world, about the cosmos, about morality and about human nature. And, of course, when the child grows up and has children of her own, she will naturally pass both classes of advice on to her own children – nonsense as well as sense – using the same solemn gravitas of manner.

On this model, we should expect that, in different geographical regions, different arbitrary beliefs having no factual basis will be handed down, to be believed with the same conviction as useful pieces of traditional wisdom, such as the belief that manure is good for the crops. We should also expect that these non-factual beliefs will evolve over generations, either by random drift or following some sort of analogue of Darwinian selection, eventually showing a pattern of significant divergence from common ancestry. Languages drift apart from a common parent, given sufficient time in geographical separation. The same is true of traditional beliefs and injunctions, handed down the generations, initially because of the programmability of the child brain.

I must again stress that the hypothesis of the programmability of the child brain is only one example of the kind of thing I mean. The message of the moths and the candle flame is more general. As a Darwinian, I am proposing a family of hypotheses, all of which have in common that they do not ask what is the survival value of religion. Instead they ask, 'What was the survival value, in the wild past, of having the kind of brain which, in the cultural present, manifests itself as religion?'* And I should add that child brains are not the only ones that are vulnerable to infection of this kind. Adult brains are too, especially if primed in childhood. Charismatic preachers can spread the word far and wide, as if they were diseased persons spreading an epidemic.

So far, the hypothesis suggests only that brains (especially child brains) are *vulnerable* to infection. It says nothing about which viruses will infect. In one sense it doesn't matter. Anything the child believes with sufficient conviction will get passed on to its children, and hence to future generations. This is a non-genetic analogue of heredity. Some people will say it is memes rather than genes. I don't want to sell memetic terminology to you today, but it is important to stress that we are not talking about genetic inheritance. What is genetically inherited, according to the theory, is the tendency of the child brain to believe what it is told. This is what makes the child brain a suitable vehicle for non-genetic heredity.

If there is non-genetic heredity, could there also be non-genetic Darwinism? Is it arbitrary which mind viruses end up exploiting the vulnerability of child brains? Or do some viruses survive better

---

* This family of hypotheses could be called 'by-product' hypotheses. Just as moth self-immolation behaviour is a by-product of the useful light compass, so religious behaviour is a by-product of – in my particular suggestion – child obedience. What else might religion be a by-product of? Another suggestion I favour is the 'vacuum gratitude' which was the topic of my Afterword to a previous essay in this section (see page 243). Gratitude is a manifestation of the beneficial reciprocation tendency of our brain. Vacuum gratitude is a by-product of this and religion is a by-product of vacuum gratitude.

than others? This is where those theories that I earlier dismissed as proximate, not ultimate, come in. If fear of death is common, the idea of immortality might survive as a mind virus better than the competing idea that death snuffs us out like a light. Conversely, the idea of posthumous punishment for sins might survive, not because children like the idea but because adults find it a useful way to control them. The important point is that survival value does not have its normal Darwinian meaning of genetic survival value. This is not the normal Darwinian conversation about why a gene survives in preference to its alleles in the gene pool. This is about why one idea survives in the pool of ideas in preference to rival ideas. It is this notion of rival ideas surviving, or failing to survive, in a pool of ideas that the word 'meme' was intended to capture.

Let's go back to first principles and remind ourselves of exactly what is going on in natural selection. The necessary condition is that accurately self-replicating information exists in alternative, competing versions. Following George C. Williams in his *Natural Selection*, I shall call them 'codices' (singular: 'codex'). The archetypal codex is a gene: not the physical molecule of DNA but the information it carries.

Biological codices, or genes, are carried around inside bodies whose qualities – phenotypes – they helped to influence. The death of the body entails the destruction of any codices that it contains, unless they have previously been passed on to another body, in reproduction. Automatically, therefore, those genes that positively affect the survival and reproduction of bodies in which they sit will come to predominate in the world, at the expense of rival genes.

A familiar example of a non-genetic codex is the so-called chain letter, although 'chain' is not a good word. It is too linear, doesn't capture the idea of explosive, exponential spread. Equally ill-named, and for the same reason, is the so-called chain reaction in an atomic bomb. Let's change 'chain letter' to 'postal virus' and look at the phenomenon through Darwinian eyes.

Suppose you received through the mail a letter which simply said: 'Make six copies of this letter and send them to six friends.' If you slavishly obeyed the instruction, and if your friends and their friends did too, the letter would spread exponentially and we'd soon be wading knee deep in letters. Of course most people would not obey such a bald, unadorned instruction. But now, suppose the letter said: 'If you do not copy this letter to six friends, you will be jinxed, a voodoo will be placed on you, and you will die young in agony.' Most people still wouldn't send it on, but a significant number probably would. Even quite a low percentage would be enough for it to take off.

The promise of reward may be more effective than the threat of punishment. We have probably all received examples of the slightly more sophisticated style of letter, which invites you to send a small sum of money to people already on the list, with the promise that you will eventually receive millions of dollars when the exponential explosion has advanced further. Whatever our personal guesses as to who might fall for these things, the fact is that many do. It is an empirical fact that chain letters circulate. No genes are involved, yet postal viruses display an entirely authentic epidemiology, including the successive waves of infection rolling around the world and including the evolution of new mutant strains of the original virus.

And the lesson for understanding religion, to repeat, is that when we ask the Darwinian question, 'What is the survival value of religion?' we don't have to mean genetic survival value. The conventional Darwinian question translates into: 'How does religion contribute to the survival and reproduction of individual religious people and hence the propagation of genetic propensities to religion?' But my point is that we don't need to bring genes into the calculation at all. There is at least something Darwinian going on here, something epidemiological going on, which has nothing to do with genes. It is the religious ideas themselves that survive, or fail to survive, in direct competition with rival religious ideas.

It is at this point that I have an argument with some of my Darwinian colleagues. Purist evolutionary psychologists will come back at me and say something like this. Cultural epidemiology is possible only because human brains have certain evolved tendencies, and by evolved we mean genetically evolved. You may document a worldwide epidemic of reverse baseball hats, or an epidemic of copycat martyrdoms, or an epidemic of total-immersion baptisms. But these non-genetic epidemics depend upon the human tendency to imitate. And we ultimately need a Darwinian – by which we mean genetic – explanation for the human tendency to imitate.

And this, of course, is where I return to my theory of childhood gullibility. I stressed that it was only an example of the kind of theory I want to propose. Ordinary genetic selection sets up childhood brains with a tendency to believe their elders. Ordinary, straight-down-the-line Darwinian selection of genes sets up brains with a tendency to imitate, hence indirectly to spread rumours, spread urban legends, and believe cock-and-bull stories in chain letters. But *given* that genetic selection has set up brains of this kind, they then provide the equivalent of a new kind of non-genetic heredity, which might form the basis for a new kind of epidemiology, and perhaps even a new kind of non-genetic Darwinian selection. I believe that religion, along with chain letters and urban legends, is one of a group of phenomena explained by this kind of non-genetic epidemiology, with the possible admixture of non-genetic Darwinian selection. If I am right, religion has no survival value for individual human beings, or for the benefit of their genes. The benefit, if there is any, is to religion itself.

Is science a religion?*

IT IS FASHIONABLE to wax apocalyptic about the threat to humanity posed by the AIDS virus, 'mad cow' disease and other infectious perils. I think a case can be made that one of the greatest of these – comparable to the smallpox virus but harder to eradicate – is faith.

Faith, being belief that isn't based on evidence, is the principal vice of any religion. And who, looking at Northern Ireland or the Middle East, can be confident that the brain virus of faith is not exceedingly dangerous? One of the stories told to young Muslim suicide bombers is that martyrdom is the quickest way to heaven – and not just heaven but a special part of heaven where they will receive their special reward of seventy-two virgin brides. It occurs to me that our best hope may be to provide a kind of 'spiritual arms control': send in specially trained theologian commandos to de-escalate the going rate in virgins.

Given the dangers of faith – and considering the accomplishments of reason and observation in the activity called science – I find it ironic that, whenever I lecture publicly, there always seems to be someone who comes forward and says: 'Of course, your science is just a religion like ours. Fundamentally, science just comes down to faith, doesn't it?'

---

* In 1996 I was honoured to be given the Humanist of the Year award by the American Humanist Association at their conference in Atlanta. This is the very slightly abridged text of my acceptance speech.

Well, science is not religion and it doesn't just come down to faith. Although it has many of religion's virtues, it has none of its vices. Science is based upon verifiable evidence. Religious faith not only lacks evidence, its independence from evidence is its pride and joy, shouted from the rooftops. Why else would Christians wax critical of Doubting Thomas? The other apostles are held up to us as exemplars of virtue because faith was enough for them. Doubting Thomas, on the other hand, required evidence. Perhaps he should be the patron saint of scientists.

One thing that provokes the comment about science being my religion is my belief in the fact of evolution. I even believe in it with passionate conviction. To some, this may superficially look like faith. But the evidence that makes me believe in evolution is not only overwhelmingly strong; it is freely available to anyone who takes the trouble to read up on it. Anyone can study the same evidence that I have studied and presumably come to the same conclusion. But if you have a belief that is based solely on faith, we can't examine your reasons. You can retreat behind the private wall of faith where we can't reach you.

Now in practice, of course, individual scientists do sometimes slip back into the vice of faith, and a few may believe so single-mindedly in a favourite theory that they occasionally falsify evidence. However, the fact that this sometimes happens doesn't alter the principle that, when they do so, they do it with shame and not with pride. The method of science is so designed that it usually finds them out in the end.

Science is actually one of the most moral, one of the most honest disciplines around – because science would completely collapse if it weren't for a scrupulous adherence to honesty in the reporting of evidence.* As James Randi has pointed out, this is one reason why scientists are so often fooled by paranormal tricksters and why the debunking role is better played by professional

---

* See the first piece in this volume, 'The values of science and the science of values'.

conjurors; scientists just don't anticipate deliberate dishonesty so well. There are other professions (no need to mention lawyers specifically) in which twisting evidence, if not falsifying it, is precisely what people are paid for and get brownie points for doing.

Science, then, is free of the main vice of religion, which is faith. But, as I pointed out, science does have some of religion's virtues. Religion may aspire to provide its followers with various benefits – among them explanation, consolation and uplift. Science, too, has something to offer in these areas.

Humans have a great hunger for explanation. It may be one of the main reasons why humanity so universally has religion, since religions do aspire to provide explanations. We come to our individual consciousness in a mysterious universe and long to understand it. Most religions offer a cosmology and a biology, a theory of life, a theory of origins and reasons for existence. In doing so, they demonstrate that religion is, in a sense, science; it's just bad science. Don't fall for the argument that religion and science operate on separate dimensions and are concerned with quite separate sorts of questions. Religions have historically always attempted to answer the questions that properly belong to science. Thus religions should not be allowed now to retreat from the ground upon which they have traditionally attempted to fight. They do offer both a cosmology and a biology. However, in both cases it is false.

Consolation is harder for science to provide. Unlike religion, science cannot offer the bereaved a glorious reunion with their loved ones in the hereafter. Those wronged on this Earth cannot, on a scientific view, anticipate a sweet comeuppance for their tormentors in a life to come. It could be argued that, if the idea of an afterlife is an illusion (as I believe it is), the consolation it offers is hollow. But that's not necessarily so; a false belief can be just as comforting as a true one, provided the believer never discovers its falsity. But if consolation comes that cheap, science can weigh in with other cheap palliatives, such as pain-killing drugs: the comfort they offer may or may not be illusory, but they do work.

Uplift, however, is where science really comes into its own. All the great religions have a place for awe, for ecstatic transport at the wonder and beauty of creation. And it's exactly this feeling of spine-shivering, breath-catching awe – almost worship – this flooding of the chest with epiphanic wonder, that modern science can provide. And it does so beyond the wildest dreams of saints and mystics. The fact that the supernatural has no place in our explanations, in our understanding of the universe and life, doesn't diminish the awe. Quite the contrary. The merest glance through a microscope at the brain of an ant or through a telescope at a long-ago galaxy of a billion worlds is enough to render poky and parochial the very psalms of praise.

Now, as I say, when it is put to me that science or some particular part of science, like evolutionary theory, is just a religion like any other, I usually deny it with indignation. But I've begun to wonder whether perhaps that's the wrong tactic. Perhaps the right tactic is to accept the charge gratefully and demand equal time for science in religious education classes. And the more I think about it, the more I realize that an excellent case could be made for this. So I want to talk a little bit about religious education and the place that science might play in it.

Among the various things that religious education might be expected to accomplish, one of its aims could be to encourage children to reflect upon the deep questions of existence, to invite them to rise above the humdrum preoccupations of ordinary life and think *sub specie aeternitatis*.

Science can offer a vision of life and the universe which, as I've already remarked, for humbling poetic inspiration far outclasses any of the mutually contradictory faiths and disappointingly recent traditions of the world's religions.

For example, how could any child in a religious education class fail to be inspired if we could get across to them some inkling of the age of the universe? Suppose that, at the moment of Christ's death, the news of it had started travelling at the maximum possible speed around the universe outwards from the Earth? How far would the terrible tidings have travelled by now? Following the theory of special relativity, the answer is that the news could not, under any circumstances whatever, have reached more than one-fiftieth of the way across one galaxy – not one-thousandth of the way to our nearest neighbouring galaxy in the hundred-million-galaxy-strong universe. The universe at large couldn't possibly be anything other than indifferent to Christ, his birth, his passion, his death. Even such momentous news as the origin of life on Earth could have travelled only across our little local cluster of galaxies. Yet so ancient was that event on our earthly timescale that, if you span its age with your open arms, the whole of human history, the whole of human culture, would fall in the dust from your fingertip at a single stroke of a nail-file.

The argument from design, an important part of the history of religion, wouldn't be ignored in my religious education classes, needless to say. The children would look at the spellbinding wonders of the living kingdoms and would consider Darwinism alongside the creationist alternatives and make up their own minds. I think the children would have no difficulty in making up their minds the right way if presented with the evidence.

It would also be interesting to teach more than one theory of creation. The dominant one in this culture happens to be the Jewish creation myth, itself based on a Babylonian creation myth. There are, of course, lots and lots of others, and perhaps they should all be given equal time (except that wouldn't leave much time for studying anything else). I understand that there are Hindus who believe that the world was created in a cosmic butter churn and Nigerian peoples who believe that the world was created by God from the excrement of ants. Don't these stories have as much right to equal time as the Judeo-Christian myth of Adam and Eve?

So much for Genesis; now let's move on to the prophets. Halley's Comet will return without fail in the year 2062. Biblical or Delphic prophecies don't begin to aspire to such accuracy; astrologers and Nostradamians dare not commit themselves to factual prognostications but, rather, disguise their charlatanry in a smokescreen of vagueness. When comets have appeared in the past, they've often been taken as portents of disaster. Astrology has played an important part in various religious traditions, including Hinduism. The three wise men were said to have been led to the cradle of Jesus by a star. We might ask the children by what physical route they imagine the alleged stellar influence on human affairs could travel.

Incidentally, there was a shocking programme on BBC radio around Christmas 1995 featuring an astronomer, a bishop and a journalist who were sent off on an assignment to retrace the steps of the three wise men. Well, you could understand the participation of the bishop and the journalist (who happened to be a religious writer) – but the astronomer was a supposedly respectable writer in that field, and yet she went along with this! All along the route, she talked about the portents of when Saturn and Jupiter were in the ascendant up Uranus or whatever it was. She doesn't actually believe in astrology, but one of the problems is that our culture has been taught to become tolerant of it, even vaguely amused by it – so much so that even scientific people who don't believe in astrology often think it's a bit of harmless fun. I take astrology very seriously indeed: I think it's deeply pernicious because it undermines rationality, and I should like to see campaigns against it.

When the religious education class turns to ethics, I don't think science actually has a lot to say, and I would replace it with rational moral philosophy. Do the children think there are absolute standards of right and wrong? And if so, where do they come from? Can you make up good working principles of right and wrong, like 'do as you would be done by' and 'the greatest good for the greatest number' (whatever that is supposed to mean)? It's a rewarding question, whatever your personal morality, to ask as an evolutionist where morals come from; by what route has the human brain gained its tendency to have ethics and morals, a feeling of right and wrong?

Should we value human life above all other life? Is there a rigid wall to be built around the species *Homo sapiens*, or should we talk about whether there are other species which are entitled to our humanistic sympathies? Should we, for example, follow the right-to-life lobby, which is wholly preoccupied with human life, and value the life of a human fetus with the faculties of a worm over the life of a thinking and feeling chimpanzee? What is the basis of this fence we erect around *Homo sapiens* – even around a small piece of fetal tissue? (Not a very sound evolutionary idea when you think about it.) When, in our evolutionary descent from our common ancestor with chimpanzees, did the fence suddenly rear itself up?

Moving on from morals to last things, to eschatology, we know from the Second Law of Thermodynamics that all complexity, all life, all laughter, all sorrow, all are hell-bent on levelling out into cold nothingness in the end. They – and we – can never be more than temporary, local buckings of the great universal slide into the abyss of uniformity. We know that the universe is expanding and will probably expand for ever, although it's possible it may contract again. We know that, whatever happens to the universe, the sun will engulf the Earth in about sixty million centuries from now.

Time itself began at a certain moment, and time may end at a certain moment – or it may not. Time may come locally to an end in miniature crunches called black holes. The laws of the universe seem to be true all over the universe. Why is this? Might the laws change in these crunches? To be really speculative, time could begin again with new laws of physics, new physical constants. And it has been plausibly suggested that there could be many universes, each one isolated so completely that, for it, the others don't exist. There might even be, as suggested by the theoretical physicist Lee Smolin, a Darwinian selection among universes.

So science could give a good account of itself in religious education. But it wouldn't be enough. I believe that some familiarity with the King James version of the Bible is important for anyone wanting to understand the allusions that appear in English literature. Together with the Book of Common Prayer, the Bible gets fifty-eight pages in the *Oxford Dictionary of Quotations*. Only Shakespeare has more. I do think that not having any kind of biblical education is unfortunate if children want to read English literature and understand the provenance of phrases like 'through a glass darkly', 'all flesh is as grass', 'the race is not to the swift', 'crying in the wilderness', 'reaping the whirlwind', 'amid the alien corn', 'eyeless in Gaza', 'Job's comforters' and 'the widow's mite'.

I want to return now to the charge that science is just a faith. The more extreme version of this charge – and one I often encounter as both a scientist and a rationalist – is an accusation of zealotry and bigotry in scientists themselves as great as that found in religious people. Sometimes there may be a little bit of justice in this accusation; but as zealous bigots, we scientists are mere amateurs at the game. We're content to argue with those who disagree with us. We don't kill them.

But I would want to deny even the lesser charge of purely verbal zealotry. There is a very, very important difference between feeling strongly, even passionately, about something because we have thought about and examined the evidence for it, and feeling strongly about something because it has been internally revealed to us, or internally revealed to somebody else in history and subsequently hallowed by tradition. There's all the difference in the world between a belief that one is prepared to defend by quoting evidence and logic and a belief that is supported by nothing more than tradition, authority or revelation. Science is founded on rational belief. Science is no religion.

Atheists for Jesus*

LIKE A GOOD recipe, the argument for a movement called 'Atheists for Jesus' needs to be built up gradually, with the ingredients mustered in advance. Start with the apparently oxymoronic title. In a society where the majority of theists are at least nominally Christian, the words 'theist' and 'Christian' are treated as near synonyms. Bertrand Russell's famous advocacy of atheism was called *Why I Am Not a Christian* rather than, as it probably should have been, *Why I Am Not a Theist*. All Christians are theists, it seems to go without saying.†

Of course Jesus was a theist, but that is the least interesting thing about him. He was a theist because, in his time, everybody

---

* This was another of my columns in *Free Inquiry*, December 2004–January 2005.

† Jews do it differently. Many people proudly call themselves Jewish atheists and observe festivals, holy days and even dietary laws. Hardly anybody self-describes as a Christian atheist, although many atheists including me lustily sing Christmas carols. Others, at least in Britain, feign religious belief and go to church as a ruse to get their children into Christian schools – because, as I documented in the 2010 Channel 4 TV programme *Faith School Menace*, they believe that faith schools tend to get good exam results. That belief is self-fulfilling because it boosts the demand to get into faith schools, and those schools consequently are able to select the best candidates for entry.

was. Atheism was not an option, even for so radical a thinker as Jesus. What was interesting and remarkable about Jesus was not the obvious fact that he believed in the god of his Jewish religion but that he rebelled against Yahweh's vengeful nastiness. At least in the teachings that are attributed to him, Jesus publicly advocated niceness and was one of the first to do so. To those steeped in the Sharia-like cruelties of Leviticus and Deuteronomy, to those brought up to fear the vindictive, Ayatollah-like god of Abraham and Isaac, a charismatic young preacher who advocated generous forgiveness must have seemed radical to the point of subversion. No wonder they nailed him.

> Ye have heard that it hath been said, An eye for an eye, and a tooth for a tooth: But I say unto you, That ye resist not evil: but whosoever shall smite thee on thy right cheek, turn to him the other also. And if any man will sue thee at the law, and take away thy coat, let him have thy cloke also. And whosoever shall compel thee to go a mile, go with him twain. Give to him that asketh thee, and from him that would borrow of thee turn not thou away. Ye have heard that it hath been said, Thou shalt love thy neighbour, and hate thine enemy. But I say unto you, Love your enemies, bless them that curse you, do good to them that hate you, and pray for them which despitefully use you, and persecute you. (Matt. 5: 38–44, King James version)

My second ingredient is another paradox and originates in my own field of Darwinism. Natural selection is a deeply nasty process. Darwin himself remarked: 'What a book a devil's chaplain might write on the clumsy, wasteful, blundering low and horridly cruel works of nature.' It was not just the facts of nature, among which he singled out the larvae of ichneumon wasps and their habit of feeding within the bodies of live caterpillars, that upset Darwin. The theory of natural selection itself seems calculated to foster selfishness at the expense of public good; violence, callous indifference to suffering, short-term greed at the expense of long-term foresight. If scientific theories could vote, Evolution would surely vote Republican.* My paradox comes from the *un*-Darwinian fact, which any of us can observe in our own circle of acquaintances, that so many individual people are kind, generous, helpful, compassionate, nice – the sort of people of whom we say, 'She's a real saint,' or, 'He's a true Good Samaritan.'

We all know people to whom we can sincerely say: 'If only everybody were like you, the world's troubles would melt away.' The milk of human kindness is only a metaphor, but, naive as it sounds, I contemplate some of my friends of both sexes, and I feel like trying to *bottle* whatever it is that makes them so kind, so selfless, so apparently un-Darwinian.

Darwinians can come up with explanations for human niceness: generalizations of the well-established models of kin selection and reciprocal altruism, the stocks-in-trade of the 'selfish gene' theory, which sets out to explain how altruism and cooperation among individual animals can flow from self-interest at the genetic level. The sort of superniceness I am talking about in humans goes too far. It is a misfiring, even a perversion of the Darwinian take on niceness. But if it is a perversion, it's the kind of perversion we need to encourage and spread.

Human superniceness is a perversion of Darwinism, because, in a wild population, it would be removed by natural selection. It is also, although I haven't the space to go into detail about this third ingredient of my recipe, an apparent perversion of the sort of rational choice theory by which economists explain human behaviour as calculated to maximize self-interest.

---

* Cynics might see this as a promising approach towards educating Republican politicians who attempt to subvert the teaching of evolution in schools. Maybe I should start with Oklahoma State Representative Todd Thomsen who, in 2009, introduced a bill in the legislature to get me banned from lecturing at the State University on the grounds (an idiosyncratic interpretation of the role of a university, to say the least) that my 'statements on the theory of evolution' were 'not representative of the thinking of the majority of the citizens of Oklahoma'.

Let's put it even more bluntly. From a rational choice point of view, or from a Darwinian point of view, human superniceness is just plain dumb. But it is the kind of dumb that should be encouraged – which is the purpose of my article. How can we do it? How shall we take the minority of supernice humans whom we all know, and increase their number, perhaps until they even become a majority in the population? Could superniceness be induced to spread like an epidemic? Could superniceness be packaged in such a form that it passes down the generations in swelling traditions of longitudinal propagation?

Well, do we know of any comparable examples, where stupid ideas have been known to spread like an epidemic? Yes, by God! *Religion*. Religious beliefs are irrational. Religious beliefs are dumb and dumber: superdumb. Religion drives otherwise sensible people into celibate monasteries or crashing into New York skyscrapers. Religion motivates people to whip their own backs, to set fire to themselves or their daughters, to denounce their own grandmothers as witches, or, in less extreme cases, simply to stand or kneel, week after week, through ceremonies of stupefying boredom. If people can be infected with such self-harming stupidity, infecting them with niceness should be a doddle.

Religious beliefs most certainly spread in epidemics and, even more obviously, they pass down the generations to form longitudinal traditions and promote enclaves of locally peculiar irrationality. We may not understand why humans behave in the weird ways we label religious, but it is a manifest fact that they do. The existence of religion is evidence that humans eagerly adopt irrational beliefs and spread them, both vertically in traditions and horizontally in epidemics of evangelism. Could this susceptibility, this palpable vulnerability to infections of irrationality, be put to genuinely good use?

Humans undoubtedly have a strong tendency to learn from and copy admired role models. Under propitious circumstances, the epidemiological consequences can be dramatic. The hairstyle of a football player, the dress sense of a singer, the speech mannerisms of a game-show host – such trivial idiosyncrasies can spread through a susceptible age-cohort like a virus. The advertising industry is professionally dedicated to the science – or it may be an art – of launching memetic epidemics and nurturing their spread. Christianity itself was spread by the equivalents of such techniques, initially by St Paul and later by priests and missionaries who systematically set out to increase the numbers of converts in what sometimes turned out to be exponential growth. Could we achieve exponential amplification of the numbers of supernice people?

I recently had a public conversation in Edinburgh with Richard Holloway, former bishop of that beautiful city. Bishop Holloway has evidently outgrown the supernaturalism that most Christians still identify with their religion (he describes himself as post-Christian and as a 'recovering Christian'). He retains a reverence for the poetry of religious myth, which is enough to keep him going to church. And, in the course of our Edinburgh discussion, he made a suggestion that went straight to my core. Borrowing a poetic myth from the worlds of mathematics and cosmology, he described humanity as a 'singularity' in evolution. He meant exactly what I have been talking about in this essay, although he expressed it differently.* The advent of human superniceness is something unprecedented in four billion years of evolutionary history. It seems likely that, after the *Homo sapiens* singularity, evolution may never be the same again.

Be under no illusions, for Bishop Holloway was not. The singularity is a product of blind evolution itself, not the creation of any unevolved intelligence. It resulted from the natural evolution of the human brain, which, under the blind forces of natural selection, expanded to the point where, all unforeseen, it overreached itself and started to behave insanely from the selfish gene's point of view. The most transparently un-Darwinian misfiring is contraception,

* He didn't mean singularity in the sense used by the trans-humanist futurist Ray Kurzweil, but was advancing a different metaphorical development of the physicists' term.

which divorces sexual pleasure from its natural function of gene propagation. More subtle overreachings include intellectual and artistic pursuits that squander, by the selfish genes' lights, time and energy that should be devoted to surviving and reproducing. The big brain achieved the evolutionarily unprecedented feat of genuine foresight: it became capable of calculating long-term consequences beyond short-term selfish gain. And, at least in some individuals, the brain overreached itself to the extent of indulging in that superniceness whose singular existence is the central paradox of my thesis. Big brains can take the driving, goal-seeking mechanisms that were originally favoured for selfish-gene reasons and divert (subvert? pervert?) them away from their Darwinian goals and into other paths.

I am no memetic engineer, and I have very little idea how to increase the numbers of the supernice and spread their memes through the meme pool. The best I can offer is what I hope may be a catchy slogan: 'Atheists for Jesus' would grace a T-shirt. There is no strong reason to choose Jesus as icon instead of some other role model from the ranks of the supernice such as Mahatma Gandhi (not the odiously self-righteous and hypocritical Mother Teresa, heavens no*). I think we owe Jesus the honour of separating his genuinely original and radical ethics from the supernatural nonsense that he inevitably espoused as a man of his time. And perhaps the oxymoronic impact of 'Atheists for Jesus' might be just what is needed to kick-start the meme of superniceness in a post-Christian society. If we play our cards right, could we lead society away from the nether regions of its Darwinian origins into the kinder and more compassionate uplands of post-singularity enlightenment?

I think a reborn Jesus would wear the T-shirt. It has become a commonplace that, were he to return today, he would be appalled at what is being done in his name by Christians ranging from the

* See Christopher Hitchens' *The Missionary Position* for substantiation of this negative judgement.

Catholic Church with its vast and ostentatious wealth to the fundamentalist religious right with its stated doctrine, explicitly contradicting Jesus, that 'God wants you to be rich'. Less obviously but still plausibly, in the light of modern scientific knowledge, I think he would see through supernaturalist obscurantism. But, of course, modesty would compel him to turn his T-shirt around to read 'Jesus for Atheists'.

AFTERWORD

This essay is worded on the assumption that Jesus was a real person who existed. There is a minority school of thought among historians that he didn't. They have a lot going for them. The gospels were written decades after Jesus' purported death, by unknown disciples who never met him but were motivated by a powerful religious agenda. More, their conception of historical fact was so different from ours that they blithely made stuff up in order to fulfil Old Testament prophecies. Matthew's virgin birth story was invented to fulfil an apparent prophecy of Isaiah which actually stemmed from a mistranslation of a Hebrew word meaning 'young woman' into a Greek word meaning 'virgin'. The earliest books of the New Testament are among the epistles, and they say almost nothing about Jesus' life, just plenty of made-up stuff about his theological significance. There's a suspicious shortage of mentions of him in any extra-biblical documents. For my purposes here I don't really care, one way or the other. If he was a fictional or mythical character, then it is that fictional character whose virtues I want us to emulate. The credit should go either to a man called Jesus or to the writer who invented him. The point of my essay remains.

But as a separate question, it is quite interesting to ask whether he really *did* exist. Jesus is the Latin form of Yehoshua, Yeshua, Yeshu, Joshua, and there were plenty of those around at the time. There were also plenty of wandering preachers, and the two sets probably overlapped. In that sense there easily could have been several Jesuses. Some of them may have suffered crucifixion: there was a lot of that about, too, in Roman times. But did any of them walk on water, turn water into wine, have a virgin mother, raise himself or anyone else from the dead, or perform any miracles that violated the laws of physics? No. Did any of them say anything so good as the Sermon on the Mount? Either one of them did, or somebody else made it up and put it into a fictional character's mouth, and that's all that matters for my essay here. Superniceness is worth spreading and religion may show us a way to spread it.

V

LIVING IN THE REAL WORLD

READING RICHARD DAWKINS on issues of public concern, whether ethics or education, law or language, can feel like plunging into a chilly sea for a swim – from the first sharp intake of breath to increasing exhilaration to emergence with a tingling sense of well-being. It has something to do, I think, with the combination of clarity of thought, felicity of expression, serious engagement with the issue at hand and sober confidence in the capacity of objective reason to offer, if not always solutions, positive paths forward in the real world.

Given the title of this section, it may seem perverse to begin with a piece that draws its own title from an ancient Greek thinker known more for his preoccupation with the ideal. But that is precisely the point. The key idea here, that of 'essentialism' or the 'tyranny of the discontinuous mind', lies in a fundamental misconception in ways of thinking about the world; in repudiating it, this essay shows how the way we think, and the way we use language, influence the ways in which we observe, analyse and understand what goes on around us. This is a masterclass in relating theoretical concept to practical experience.

Among the targets of that essay are 'finger-wagging, hectoring lawyers' demanding yes-or-no answers to complex questions about risk, safety, guilt. The legal system comes in for more criticism in the second piece, ' "Beyond reasonable doubt"?', which interrogates the practice of trial by jury with a forensic rigour most barristers would be proud to marshal.

'But can they suffer?' tackles the conundrum of pain, and our human perceptions of it in ourselves and in other beings. It challenges the widely held 'speciesist' assumption that privileges human experience over that of other animals, and offers good reason to doubt that there is any correlation between mental capacity and the capacity to suffer pain. 'I love fireworks, but . . .' brings the theme of non-human distress closer to home in calling for more consideration of how pets and wild animals – not to mention war veterans – experience the explosive noise that is part of so many displays.

The next piece, 'Who would rally against reason?', issues a rousing invitation to the Reason Rally in Washington DC, beginning with a hymn to the achievements of reason and ending with another call to arms in its defence. If that piece leaves some British readers feeling slightly smug, the next, 'In praise of subtitles; or, a drubbing for dubbing', should banish any such self-satisfaction in a great many of us who listen in awe to the fluency of Europeans speaking English. This is more than a bewailing of a national shortcoming: harnessing scientific imagination to real-world observation, it suggests reasons beyond laziness or the long imperial shadow and makes fascinating proposals for redress.

So many problems to be tackled, so many obstacles in the way: no wonder a writer of intellectual heft, imaginative reach and strong public engagement gets frustrated at times. The final piece in this section gives just one small glimpse of what might happen *if* Richard Dawkins ruled the world . . .

G.S.

The dead hand of Plato*

WHAT PERCENTAGE OF the British population lives below the poverty line? When I call that a silly question, a question that doesn't deserve an answer, I'm not being callous or unfeeling about poverty. I care very much if children starve or pensioners shiver with cold. My objection – and this is just one of many examples – is to the very idea of a line: a gratuitously manufactured discontinuity in a continuous reality.

Who decides how poor is poor enough to qualify as below the 'poverty line'? What is to stop us moving the line and thereby changing the score? Poverty/wealth is a continuously distributed quantity, which might be measured as, say, income per week. Why throw away most of the information by splitting a continuous variable into two discontinuous categories: above and below the 'line'? How many of us lie below the stupidity line? How many runners exceed the fast line? How many Oxford undergraduates lie above the first-class line?

Yes, we in universities do it too. Examination performance, like most measures of human ability or achievement, is a continuous

---

* I was invited to be guest editor of *New Statesman* for the Christmas double issue of 2011. This article is largely taken from 'The tyranny of the discontinuous mind', which was my essay in that issue, but also incorporates parts of my chapter, 'Essentialism', in John Brockman's edited volume *This Idea Must Die: scientific theories that are blocking progress.*

variable, whose frequency distribution is bell-shaped. Yet British universities insist on publishing a class list, in which a minority of students receive first-class degrees, rather a lot obtain seconds (nowadays subdivided into upper and lower seconds), and a few get thirds. That might make sense if the distribution had three or four peaks with deep valleys in between, but it doesn't. Anybody who has ever marked an exam knows that the bottom of one class is separated from the top of the class below by a small fraction of the distance that separates it from the top of its own class. This fact alone points to a deep unfairness in the system of discontinuous classification.

Examiners go to great trouble to assign a score, perhaps out of 100, to each exam script. Scripts are double- or even triple-marked by different examiners, who may then argue the nuances of whether an answer deserves 55 or 52 marks. Marks are scrupulously added up, normalized, transformed, juggled and fought over. The final marks that emerge, and the rank orders of students, are as richly informative as conscientious examiners can achieve. But then what happens to all that richness of information? Most of it is thrown away, in reckless disregard for all the labour and nuanced deliberation and adjusting that went into the marking process. The students are bundled into three or four discrete classes, and that is all the information that penetrates outside the examiners' room.

Cambridge mathematicians, as one might expect, finesse the discontinuity and leak the rank order. It became informally known that Jacob Bronowski was the 'Senior Wrangler' of his year, Bertrand Russell the Seventh Wrangler of his year and so on. At other universities, too, tutors' testimonials may say things like, 'Not only did she get a first: I can tell you in confidence that the examiners ranked her number 3 of her entire class of 106 in the university.' That is the kind of information that really counts in a letter of recommendation. And it is that very information that is wantonly thrown away in the officially published class list.

Perhaps such wastage of information is inevitable: a necessary evil. I don't want to make too much of it. What is more serious is that there are some educators – dare I say especially in non-scientific subjects – who fool themselves into believing that there is a kind of Platonic ideal called the 'First Class Mind' or 'Alpha Mind': a qualitatively distinct category, as distinct as female is from male, or sheep from goat. This is an extreme form of what I am calling the discontinuous mind. It can probably be traced to the 'essentialism' of Plato – one of the most pernicious ideas in all history.

Plato took his characteristically Greek geometer's view of things and forced it where it didn't belong. For Plato, a circle, or a right triangle, was an ideal form, definable mathematically but never realized in practice. A circle drawn in the sand was an imperfect approximation to the ideal Platonic circle hanging in some abstract space. That works for geometric shapes like circles; but essentialism has been applied to living things, and Ernst Mayr blamed this for humanity's late discovery of evolution – as late as the nineteenth century. If you treat all flesh-and-blood rabbits as imperfect approximations to an ideal Platonic rabbit, it won't occur to you that rabbits might have evolved from a non-rabbit ancestor, and might evolve into a non-rabbit descendant. If you think, following the dictionary definition of essentialism, that the *essence* of rabbitness is 'prior to' the *existence* of rabbits (whatever 'prior to' might mean, and that's a nonsense in itself) evolution is not an idea that will spring readily to your mind, and you may resist when somebody else suggests it.

For legal purposes, say in deciding who can vote in elections, we need to draw a line between adult and non-adult. We may dispute the rival merits of eighteen versus twenty-one or sixteen, but everybody accepts that there has to be a line, and the line must be a birthday. Few would deny that some fifteen-year-olds are better qualified to vote than some forty-year-olds. But we recoil from the voting equivalent of a driving test, so we accept the age line as a necessary evil. But perhaps there are other examples where we should be less willing to do so. Are there cases where the tyranny of the discontinuous mind leads to real harm: cases where we should actively rebel against it? Yes.

Essentialism bedevils moral controversies such as those over abortion and euthanasia. At what point is a brain-dead accident victim defined as 'dead'? At what moment during development does an embryo become a 'person'? Only a mind infected with essentialism would ask such questions. An embryo develops gradually from single-celled zygote to newborn baby, and there's no single instant when 'personhood' should be deemed to have arrived. The world is divided into those who get this truth and those who wail: 'But there has to be *some* moment when the fetus becomes human.' No, there really doesn't, any more than there has to be a day when a middle-aged person becomes old. It would be better – though still not ideal – to say the embryo goes through stages of being a quarter human, half human, three-quarters human . . . The essentialist mind will recoil from such language and accuse me of all manner of horrors for denying the *essence* of humanness.

There are those who cannot distinguish a sixteen-cell embryo from a baby. They call abortion murder, and feel righteously justified in committing real murder against a doctor – a thinking, feeling, sentient adult, with a loving family to mourn him. The discontinuous mind is blind to intermediates. An embryo is either human or it isn't. Everything is this or that, yes or no, black or white. But reality isn't like that.

For purposes of legal clarity, just as the eighteenth birthday is defined as the moment of getting the vote, it may be necessary to draw a line at some arbitrary moment in embryonic development after which abortion is prohibited. But personhood doesn't spring into existence at any one moment: it matures gradually, and it goes on maturing through childhood and beyond.

To the discontinuous mind, an entity either is a person or is not. The discontinuous mind cannot grasp the idea of a half person, or a three-quarters person. Some absolutists go right back to conception as the moment when the person comes into existence – the instant the soul is injected – so all abortion is murder by definition. The Catholic Doctrine of the Faith entitled *Donum Vitae* says:

> From the time that the ovum is fertilized, a new life is begun which is neither that of the father nor of the mother; it is rather the life of a new human being with his own growth. It would never be made human if it were not human already. To this perpetual evidence . . . modern genetic science brings valuable confirmation. It has demonstrated that, from the first instant, the program is fixed as to what this living being will be: a man, this individual-man with his characteristic aspects already well determined. Right from fertilization is begun the adventure of a human life . . . *

It is amusing to tease such absolutists by confronting them with a pair of identical twins (they split after fertilization, of course) and asking which twin got the soul, which twin is the non-person: the zombie. A puerile taunt? Maybe. But it hits home because the belief that it destroys is puerile. And ignorant.

'It would never be made human if it were not human already.' Really? Are you serious? Nothing can become something if it is not that something already? Is an acorn an oak tree? Is a hurricane the barely perceptible zephyr that seeds it? Would you apply your doctrine to evolution too? Do you suppose there was a moment in evolutionary history when a non-person gave birth to the first person?

Paleontologists will argue passionately about whether a particular fossil is, say, *Australopithecus* or *Homo*. But any evolutionist knows there must have existed individuals who were exactly intermediate. It's essentialist folly to insist on shoehorning your fossil into one genus or the other. There never was an *Australopithecus* mother who gave birth to a *Homo* child, for every child ever born belonged to the same species as its mother. The whole system of labelling species with discontinuous names is geared to a time slice, such as the present, in which ancestors have been conveniently

---

* To quote the immortal words of Monty Python's Michael Palin, 'You're a Catholic the moment Dad came.'

expunged from our awareness. If by some miracle every ancestor were preserved as a fossil, discontinuous naming would be impossible.* Creationists are misguidedly fond of citing 'gaps' as embarrassing for evolutionists, but gaps are a fortuitous boon for taxonomists who, with good reason, want to give species discrete names. Quarrelling about whether a fossil is 'really' *Australopithecus* or *Homo* is like quarrelling over whether George should be called 'tall'. He's five foot ten, doesn't that tell you what you need to know?

If a time machine could serve up to you your 200-million-greats-grandfather, you would eat him with *sauce tartare* and a slice of lemon. He was a fish. Yet you are connected to him by an unbroken line of intermediate ancestors, every one of whom belonged to the same species as its parents and its children.

'I've danced with a man who's danced with a girl who's danced with the Prince of Wales,' as the song goes. I could mate with a woman, who could mate with a man, who could mate with a woman who ... after a sufficient number of steps ... could mate with an ancestral fish, and produce fertile offspring. To invoke our time machine again, you could not mate with *Australopithecus* (at least, the pairing would not produce fertile offspring) but you are connected to *Australopithecus* by an unbroken chain of intermediates who could interbreed with their neighbours in the chain every step of the way. And the chain goes on backwards, unbroken, to that fish of the Devonian period and beyond. But for the extinction of the intermediates which connect humans to the ancestor we share with pigs (it pursued its shrew-like existence eighty-five million years ago in the shadow of the dinosaurs), and but for the extinction of the intermediates that connect the same ancestor to modern pigs, there would be no clear separation between *Homo sapiens* and *Sus scrofa*. You could breed with X who could breed with Y who could breed with (... fill in several thousand intermediates ...) who could produce fertile offspring by mating with a sow.

---

* Walking around would be pretty difficult, too: we'd be tripping over fossils every step of the way.

It is only the discontinuous mind that insists on drawing a hard and fast line between a species and the ancestral species that birthed it. Evolutionary change is gradual: there never was a line between any species and its evolutionary precursor.*

In a few cases the intermediates have failed to go extinct, and the discontinuous mind really is faced with the problem in stark reality. Herring gulls (*Larus argentatus*) and lesser black-backed gulls (*Larus fuscus*) breed in mixed colonies in western Europe and don't interbreed. This defines them as good, separate species. But if you travel in a westerly direction around the northern hemisphere and sample the gulls as you go, you find that the local gulls vary from the light grey of the herring gull, getting gradually darker as you progress around the North Pole, until eventually, when you go all the way round to western Europe again, they have darkened so far that they 'become' lesser black-backed gulls. What's more, the neighbouring populations interbreed with each other all the way around the ring, even though the ends of the ring, the two species we see in Britain, don't interbreed. Are they distinct species or not? Only those tyrannized by the discontinuous mind feel obliged to answer that question. If it were not for the accidental extinction of evolutionary intermediates, every species would be linked to every other by interbreeding chains like these gulls.

Essentialism rears its ugly head in racial terminology. The majority of 'African Americans' are of mixed race. Yet so entrenched is our essentialist mindset, American official forms require everyone to tick one race/ethnicity box or another: no room for intermediates. In the United States today, a person will be called 'African American' even if only, say, one of his eight great-grandparents was of African descent.

Colin Powell and Barack Obama are described as black. They do have black ancestors, but they also have white ancestors, so why

---

* There are some exceptions, especially in plants, where a new species, defined by the criterion of inability to interbreed, comes into being in a single generation.

don't we call them white? It is a weird convention that the descriptor 'black' behaves as the cultural equivalent of a genetic dominant. Gregor Mendel, the father of genetics, crossed wrinkled and smooth peas and the offspring were all smooth: smoothness is 'dominant'. When a white person breeds with a black person the child is intermediate but is labelled 'black': the cultural label is transmitted down the generations like a dominant gene, and this persists even to cases where, say, only one out of eight great-grandparents was black and it may not show in skin colour at all. It is the racist 'contamination metaphor' (as Lionel Tiger pointed out to me), the 'touch of the tarbrush'. Our language lacks the equivalent 'touch of whitewash' and is ill-equipped to deal with a continuum of intermediates. Just as people must lie below or above the poverty 'line', so we classify people as 'black' even if they are in fact intermediate. When an official form invites us to tick a 'race' or 'ethnicity' box I recommend crossing it out and writing 'human'.

In US presidential elections every state (except Maine and Nebraska) has to end up labelled either Democrat or Republican, no matter how evenly divided the voters in that state might be. Each state sends to the Electoral College a number of delegates which is proportional to the population of the state. So far so good. But the discontinuous mind insists that all the delegates from a given state have to vote the same way. This 'winner-take-all' system was shown up in all its fatuity in the 2000 election when there was a dead heat in Florida. Al Gore and George Bush received the same number of votes as each other, the tiny, disputed difference being well within the margin of error. Florida sends twenty-five delegates to the Electoral College.* The Supreme Court was asked to decide which candidate should receive all twenty-five votes (and therefore the presidency). Since it was a dead heat, it might have seemed reasonable to allot thirteen votes to one candidate and twelve to the other. It would have made no difference whether Bush or Gore received the thirteen votes: either way Gore would have been

⊩   * That was the number in 2000. It varies from year to year.

President. Indeed, Gore could have given Bush twenty-two of the twenty-five Electoral College delegates and still won the presidency.

I am not saying the Supreme Court should actually have split the Florida delegates. They had to abide by the rules, no matter how idiotic. I would say that, given the lamentable constitutional rule that the twenty-five votes had to be bound together as a one-party block, natural justice should have led the court to allocate the twenty-five votes to the candidate who would have won the election if the Florida delegates had been divided, namely Gore. But that is not the point I am making here. My point here is that the winner-take-all idea of an Electoral College in which each state has an indivisible block of members, either all Democrat or all Republican no matter how close the vote, is a shockingly undemocratic manifestation of the tyranny of the discontinuous mind. Why is it so hard to admit that there are intermediates, as Maine and Nebraska do? Most states are not 'red' or 'blue' but a complex mixture.*

---

* If the Electoral College were ever to be abolished it would have to be done by constitutional amendment, and that is difficult. It requires a two-thirds majority in both houses of Congress and must be ratified by three-quarters of the state legislatures. The worst of both worlds would be a piecemeal reform by one state or another, following the example of Maine and Nebraska and allocating Electoral College votes pro rata. An idealistic, but probably unworkable alternative would be to revert to a true Electoral College as it was originally conceived. This would be like the College of Cardinals that chooses a new Pope, except that the members of the Electoral College would be elected, not appointed: a body of respected citizens, elected by the voters, who would meet to evaluate all the (potentially many) candidates for President – take up references, read their publications, interview them, vet them for security and health, and finally vote and announce their choice to the world with a puff of smoke: *habemus praesidem*. That's something like the way the US Electoral College actually started. The rot set in when delegates to the College became mere ciphers, pledged to support particular presidential

Scientists are called upon by governments, by courts of law, and by the public at large, to give a definite, absolute, yes-or-no answer to important questions, for example questions of risk. Whether it's a new medicine, a new weedkiller, a new power station or a new airliner, the scientific 'expert' is peremptorily asked: Is it safe? Answer the question! Yes or no? Vainly the scientist tries to explain that safety and risk are not absolutes. Some things are safer than others, and nothing is perfectly safe. There is a sliding scale of intermediates and probabilities, not hard-and-fast discontinuities between safe and unsafe. That is another story and I have run out of space.

But I hope I have said enough to suggest that the summary demand for an absolute yes-or-no answer, so beloved of journalists, politicians and finger-wagging, hectoring lawyers, is yet another unreasonable expression of a kind of tyranny, the tyranny of the discontinuous mind, the dead hand of Plato.

---

candidates. Unfortunately, my version probably couldn't work, if only because it would be vulnerable to corruption, and pledging in advance would probably creep back in.

'Beyond reasonable doubt'?*

In a court of law – say, at a murder trial – a jury is asked to decide, beyond reasonable doubt, whether a person is guilty or not guilty. In several jurisdictions, including thirty-four states of the US, a guilty verdict may result in an execution. Numerous cases are on record where later evidence not available at the time of trial, especially DNA evidence, has retrospectively reversed an old verdict and in some cases led to a posthumous pardon.

Courtroom dramas accurately portray the suspense that hangs in the air when the jury returns to deliver its verdict. All, including the lawyers on both sides and the judge, hold their breath while they wait to hear the foreman of the jury pronounce the words 'guilty' or 'not guilty'. However, if the phrase 'beyond reasonable doubt' means what it says, there should be no doubt of the outcome in the mind of anybody who has sat through the same trial as the jury. That includes the judge who, as soon as the jury has delivered its verdict, is prepared to give the order for execution – or to release the prisoner without a stain on his character.

And yet, before the jury returned, there was enough 'reasonable

---

* I have no training in law, as will no doubt be apparent to those who have. But I've served on three juries where I was instructed that proof of guilt must be established 'beyond reasonable doubt'. The meaning of 'reasonable doubt' is something upon which a scientist might have something to say. This was what I said in the *New Statesman* of 23 January 2012.

doubt' in that same judge's mind to keep him on tenterhooks waiting for the verdict.

You cannot have it both ways. Either the verdict is beyond reasonable doubt, in which case there should be no suspense while the jury is out; or there is real, nail-biting suspense, in which case you cannot claim that the facts have been proved 'beyond reasonable doubt'.

American weather forecasters deliver probabilities, not certainties: 'Eighty per cent probability of rain.' Juries are not allowed to do that, but it's what I felt like doing when I served on one. 'What is your verdict, guilty or not guilty?' 'Seventy-five per cent probability of guilt, m'lud.' That would be anathema to our judges and lawyers. There must be no shades of grey: the system insists on certainty, yes or no, guilty or not guilty. Judges may refuse even to accept a divided jury and will send members back into the jury room with instructions not to emerge again until they have somehow managed to achieve unanimity. How is that 'beyond reasonable doubt'?

In science, for an experiment to be taken seriously, it must be repeatable. Not all experiments are repeated – we have not world enough and time* – but controversial results must be repeatable, or we don't have to believe them. That is why the world of physics waited for repeat experiments before taking up the claim that neutrinos can travel faster than light – and indeed the claim was eventually rejected.

Shouldn't the decision to execute somebody, or imprison them for life, be taken seriously enough to warrant a repeat of the experiment? I'm not talking about a retrial. Nor an appeal, although that is desirable, and happens when there is some disputed point of law or new evidence. But suppose every trial had two juries, sitting in the same courtroom yet forbidden to talk to each other. Who

---

* Andrew Marvell's context was different but his lament works here too.

will bet that they would always reach the same verdict? Does *anybody* think a second jury is likely to have acquitted O. J. Simpson?

My guess is that, if the two-jury experiment were run over a large number of trials, the frequency with which two groups would agree on a verdict would run at slightly higher than 50 per cent. But anything short of 100 per cent makes one wonder at the 'beyond reasonable doubt' held to be sufficient to send somebody to the electric chair. And would anybody bet on 100 per cent concordance between two juries?

Isn't it enough, you may say, that there are twelve people on the jury? Doesn't that provide the equivalent of twelve replications of the experiment? No, it doesn't, because the twelve jurors are not independent of one another; they are locked in a room together.

Anybody who has ever been on a jury (I've been on three) knows that authoritative and articulate speakers sway the rest. *Twelve Angry Men* is fiction and doubtless exaggerated, but the principle remains. A second jury without the Henry Fonda character would have found the boy guilty. Should a death sentence depend on the lucky break of whether a particularly perceptive or persuasive individual happens to be picked for jury duty?

I am not suggesting that we should introduce a two-jury system in practice. I suspect that two independent juries of six people would produce a fairer result than a single jury of twelve, but what would you do on those many (as I suspect) cases where the two juries disagreed? Would the two-jury system amount to a bias in favour of the defence? I can't suggest any well-worked-out alternative to the present jury system, but I still think it is terrible.

I strongly suspect that two judges, forbidden to talk to each other, would have a higher concordance rate than two juries and might even approach 100 per cent. Yet that, too, is open to the objection that the judges are likely to be drawn from the same class of society as each other and to be of similar age, and might consequently share the same prejudices.

What I am proposing, as a bare minimum, is that we should acknowledge that 'beyond reasonable doubt' is a hollow and empty phrase. If you defend the single-jury system as delivering a verdict 'beyond reasonable doubt', you are committed to the strong view, whether you like it or not, that two juries would always produce the same verdict. And when you put it like that, will *anybody* stand up and bet on 100 per cent concordance?

If you place such a bet, you are as good as saying that you wouldn't bother to stay in court to hear the verdict, because the verdict should be obvious to anybody who had sat through the trial, including the judge and the lawyers on both sides. No suspense. No tenterhooks.

There may be no practical alternative, but let's not pretend. Our courtroom procedures make a mockery of 'beyond reasonable doubt'.

But can they suffer?*

T HE GREAT MORAL philosopher Jeremy Bentham, founder of utilitarianism, famously said: 'The question is not, "Can they reason?" nor, "Can they talk?" but rather, "Can they suffer?"' Most people get the point, but they treat *human* pain as especially worrying because they vaguely think it sort of obvious that a species' ability to suffer must be positively correlated with its intellectual capacity. Plants cannot think, and you'd have to be pretty eccentric to believe they can suffer. Plausibly the same might be true of earthworms. But what about cows?

And what about dogs? I find it almost impossible to believe that René Descartes, not known as a monster, carried his philosophical belief that only humans have minds to such a confident extreme that he would blithely spreadeagle a live mammal on a board and dissect it. You'd think that, in spite of his philosophical reasoning, he might have given the animal the benefit of the doubt. But he stood in a long tradition of vivisectionists including Galen and Vesalius, and he was followed by William Harvey and many others.

How could they bear to do it: tie a struggling, screaming mammal down with ropes and dissect its living heart, for example? Presumably they believed what came to be articulated by Descartes: that non-human animals have no soul and feel no pain.

Most of us nowadays believe that dogs and other non-human

* First published on boingboing.net in 2011.

mammals can feel pain, and no reputable scientist today would follow Descartes' and Harvey's horrific example and dissect a living mammal without anaesthetic. British law, among others, would severely punish them if they did (although invertebrates are not so well protected, not even large-brained octopuses). Nevertheless, most of us seem to assume, without question, that the capacity to feel pain is positively correlated with mental capability – with the ability to reason, think, reflect and so on. My purpose here is to question that assumption. I see no reason at all why there should be a positive correlation. Pain feels primal, like the ability to see colour or hear sounds. It feels like the sort of sensation you don't need intellect to experience. Feelings carry no weight in science but, at the very least, shouldn't we give the animals the benefit of the doubt?

Without going into the interesting literature on animal suffering (see, for instance, Marian Stamp Dawkins' excellent book of that title, and her subsequent *Why Animals Matter*), I can see a Darwinian reason why there might even be a negative correlation between intellect and susceptibility to pain. I approach this by asking what, in the Darwinian sense, pain is for. It is a warning not to repeat actions that tend to cause bodily harm. Don't stub your toe again, don't tease a snake or sit on a hornet, don't pick up embers however prettily they glow, be careful not to bite your tongue. Plants have no nervous system capable of learning not to repeat damaging actions, which is why we cut live lettuces without compunction.

It is an interesting question, incidentally, why pain has to be so damned painful. Why not equip the brain with the equivalent of a little red flag, painlessly raised to warn, 'Don't do that again'? In *The Greatest Show on Earth*, I suggested that the brain might be torn between conflicting urges and tempted to 'rebel', perhaps hedonistically, against pursuing the best interests of the individual's genetic fitness, in which case it might need to be whipped agonizingly into line. I'll let that pass and return to my primary question for today: would you expect a positive or a negative correlation between mental ability and ability to feel pain? Most people unthinkingly assume a positive correlation, but why?

Isn't it plausible that a clever species such as our own might need less pain, precisely because we are capable of learning faster, or intelligently working out what is good for us and what damaging events we should avoid? Isn't it plausible that an unintelligent species might need a massive wallop of pain, to drive home a lesson that we can learn with less powerful inducement?

At very least, I conclude that we have no general reason to think that non-human animals feel pain less acutely than we do, and we should in any case give them the benefit of the doubt. Practices such as branding cattle, castration without anaesthetic, and bullfighting should be treated as morally equivalent to doing the same thing to human beings.

I love fireworks, but . . .

On 12 October 1984, a Provisional IRA member planted a bomb in the Grand Hotel, Brighton, in an attempt to assassinate the Prime Minister. That objective failed, although five people were killed and many injured. Would we want a national festival every 12 October when we all let off fireworks to commemorate this event? And if, in addition, we burned the perpetrator, Patrick Magee, in effigy all over the country, wouldn't our revulsion be increased?

Bonfire Night, with its 'remember, remember' fireworks, commemorates a mass assassination attempt in 1605.* A terrorist

---

* Non-British readers will need to know that the 'gunpowder plot' of 5 November 1605 was a Catholic plan to blow up Parliament and the Protestant King James I. A zealous Catholic convert, Guy Fawkes, was arrested guarding the barrels of gunpowder on the eve of the planned explosion. Every year on 5 November to this day, large bonfires are lit all over Britain, a 'guy' (a stuffed cloth effigy of a moustached man wearing a tall hat) is burned on top of each bonfire, and fireworks are let off. In the weeks before 'Bonfire Night' children traditionally parade their 'guy' around the streets, begging for money with which to buy fireworks: 'Penny for the Guy, Mister?' (although nowadays a penny wouldn't buy many fireworks). Most British children are capable of reciting a nursery rhyme that begins, 'Remember, remember the fifth of November, gunpowder, treason and plot.' I didn't know the rest of the rhyme, so I looked it up. It includes the lines, 'A rope, a rope, to hang the Pope; A penn'orth of cheese to choke him. A pint of beer to wash

bomb plot, even a failed one, sounds a pretty nasty thing to celebrate, which was of course why I brought up the comparison with the Brighton hotel plot. But Guy Fawkes is separated from us by more than four hundred years: long enough for the commemoration to suggest not bad taste but the quaintness of distant history. So I'm not trying to be a killjoy, a November Scrooge.

And I do love fireworks. Always have. For me the appeal is to the eye more than the ear – the spectacular colours that psychedelically paint the sky, flares lighting the smiling faces of children waving sparklers, the whirring of Catherine Wheels (again, historical distance helps us forget that that name has a pretty nasty provenance too). I don't so much get the appeal of loud bangs, but presumably some people love them or the manufacturers wouldn't put them in. So I don't want to deny that fireworks, even the bangs, are fun, and I have much enjoyed Bonfire Night over the years, from childhood on.

But although I love fireworks I also love animals. Including human animals, but just now I'm talking about non-human animals. Like our little dogs, Tycho and Cuba, who are only two among millions all over the country that are terrorized every year by the prodigiously anti-social decibels of modern fireworks. It would be tolerable if it happened only on 5 November. But over the years 'November the Fifth' has expanded relentlessly in both directions.* It seems that many people, having bought their fireworks, are too impatient to wait until the night itself. Or they

---

it down, And a jolly good fire to burn him.' The Protestant enmity in the rhyme finds its echoes today in the slogans of the Orangemen of Northern Ireland, but nowadays we have to use the euphemisms Loyalists and Nationalists instead of Protestants and Catholics. Religion cannot be admitted as the motivation for murder. A version of this article was published in the *Daily Mail* on the eve of Guy Fawkes' Day, 4 November 2014.

* I'm told that the same 'spreading' happens in America around the Fourth of July.

enjoyed the night itself so much that they can't resist reprising it week after week thereafter. And in Oxford the firework season is not a limited season at all but extends to most weekends throughout the university terms.

If it were only Tycho and Cuba whose lives are made a misery, I'd shut up about it. But when I tweeted my misgivings about the noise, the response from other owners of dogs, cats and horses was overwhelming. This subjective impression is confirmed by scientific studies. The veterinary literature lists more than twenty physiologically measurable symptoms of distress in dogs resulting from fireworks. In extreme cases the fear caused by fireworks has even led to normally gentle dogs biting their owners. It's estimated that some 50 per cent of dogs and 60 per cent of cats suffer from firework-phobia.

Then think about all the wild animals all over the country. And cattle, pigs and other farm livestock. There is no reason to believe wild animals, whom we don't see, are any less terrified than domestic pets whom we do. Rather the reverse, when you consider that loved pets like Tycho and Cuba have human comforters to soothe and console them. Wild animals suddenly, without warning, have their natural environment and peaceful nights polluted by the acoustic equivalent of a First World War battle. Talking of which, among the sympathetic responders to my firework tweets were war veterans suffering from the modern equivalent of First World War shell-shock.

What should be done? I wouldn't call for a total ban on fireworks (as enforced in some jurisdictions, including Northern Ireland during the Troubles*). Two compromises are commonly suggested. First, fireworks might be restricted to certain special days in a year, such as Guy Fawkes' Night and New Year's Eve. Other special occasions – big parties, balls and the like – could be accommodated by individual applications, along the same lines as permissions to

* Because the police couldn't hear the difference between loud fireworks and bombs.

play loud music on special occasions. The other suggested compromise is to allow firework displays to be put on by public bodies but not any old private citizens in their own back gardens. I would suggest a third compromise, which might reduce the need for the other two: allow visually appealing fireworks but put a severe restriction on noise. Quiet fireworks do exist.

Although the replies to my tweets were overwhelmingly in agreement, there were two dissenting strands, which need to be taken seriously. First, wouldn't a legal restriction on fireworks infringe personal liberties? And second, shouldn't the pleasure of humans have priority over the feelings of 'mere animals'?

The personal liberties point is superficially persuasive. Several tweeters said that what people do in their own gardens – on their own private property – is their own business and nobody else's, especially not the business of the 'nanny state'. But the sound and shock waves from a loud explosion radiate outwards far beyond the boundaries of anybody's garden. Neighbours who don't like the flashes and colours of fireworks can block them by drawing the curtains. No such blocks are effective against loud bangs. Noise pollution is antisocial in a peculiarly inescapable way, which is why the Noise Abatement Society is so necessary.

What about the 'mere animals' plea? Isn't human pleasure more important than terrified dogs, cats, horses, cows, rabbits, mice, weasels, badgers and birds? The presumption that humans matter more than other animals lies deep within us. It's a difficult philosophical problem, and this is not the place to go into it in depth. Just a couple of thoughts.

First, although the reasoning power and intelligence of non-human animals is far inferior to ours, the ability to suffer – feel pain or fear – doesn't depend on reasoning or intelligence.* An Einstein is no more capable of feeling pain or fear than a Sarah Palin. And there's no obvious reason to suppose that a dog or a badger is less capable of suffering pain or fear than any human.

‖   * As I argued in the previous article in this collection.

In the case of fear of fireworks, there might even be reason to think the opposite. Humans understand what fireworks are. Human children can be consoled with a verbal explanation: 'It's OK, darling, they're only fireworks, they're fun, nothing to worry about.' You can't do that with non-human animals.

Let's not be killjoys. But fireworks are nearly as appealing if silent. And our present disregard of millions of sentient beings incapable of understanding what fireworks are but fully capable of dreading them, is utterly – albeit usually unwittingly – selfish.

AFTERWORD

I hope this essay doesn't come across as too parochially British. It is only incidentally about Guy Fawkes' Night. Fireworks pollute the sound waves of countries throughout the world, often in celebration of particular days such as America's Fourth of July, or festivals such as the Hindu Diwali or Chinese New Year. And animals the world over are uncomprehending and terrorized.

Who would rally against reason?*

HOW HAVE WE come to the point where reason needs a rally to defend it? To base your life on reason means to base it on evidence and logic. Evidence is the only way we know to discover what's true about the real world. Logic is how we deduce the consequences that follow from evidence. Who could be against either? Alas, plenty of people, which is why we need the Reason Rally.

Reason, as played out in the grand cooperative enterprise called science, makes me proud of *Homo sapiens*. *Sapiens* literally means 'wise', but we have deserved the accolade only since we crawled from the swamp of primitive superstition and supernatural gullibility and embraced reason, logic, science and evidence-based truth.

We now know the age of our universe (thirteen to fourteen billion years), the age of the Earth (four to five billion years), what

* The Reason Rally, in the National Mall, Washington DC, was first held on 24 March 2012, and I published the original version of this essay in the *Washington Post* to encourage people to attend. The rally was a great success. An estimated thirty thousand people stood, in the pouring rain, to hear speakers and entertainers, scientists and musicians. Four years later there was a repeat performance in the same massively impressive venue. I unfortunately had to miss it for health reasons, but I published (on RichardDawkins.net, 31 May 2016) a revised version of my *Washington Post* rallying cry, and it is this updated version that is reproduced here.

we and all other objects are made of (atoms), where we come from (evolved from other species), why all species are so well adapted to their environments (natural selection of their DNA). We know why we have night and day (Earth spins like a top), why we have winter and summer (Earth is tilted), what is the maximum speed at which anything can travel (two-thirds of a billion miles per hour). We know what the sun is (one star among billions in the Milky Way galaxy), we know what the Milky Way is (one galaxy among billions in our universe). We understand what causes smallpox (a virus, which we have eradicated), polio (a virus, which we have nearly eradicated), malaria (a protozoan, still here but we're working on it), syphilis, tuberculosis, gangrene, cholera (bacteria, and we know how to kill them). We have built planes that can cross the Atlantic in hours, rockets that safely land men on the moon and robot vehicles on Mars, and might one day save our planet by diverting a meteor of the kind that – we now understand – killed the dinosaurs.* Thanks to evidence-based reason we are blessedly liberated from ancient fears of ghosts and devils, evil spirits and djinns, magic spells and witches' curses.

Who then would rally against reason? The following statements will sound all too familiar.

'I don't trust educated intellectuals, elitists who know more than I do. I'd prefer to vote for somebody like me, rather than somebody who is actually qualified to be President.'

What other than this mentality accounts for the popularity of Donald Trump, Sarah Palin, George W. Bush – politicians who flaunt their ignorance as a vote-winning virtue?[†] You want your

---

* See my introduction to this volume.

† In the 2016 UK referendum, prominent politicians leading the Leave Europe campaign fired off remarks such as 'I think people

airline pilot to be educated in aeronautics and navigation. You want your surgeon to be learned in anatomy. Yet when you vote for a President to lead a great country, you prefer somebody who is ignorant and proud of it, someone you'd enjoy having a drink with, rather than somebody qualified for high office? If you are such a voter, you will not join the Reason Rally.

'Rather than have them learn modern science, I'd prefer my children to study a book written in 800 BC by unidentified authors whose knowledge and qualifications were of their time. If I can't trust the school to shield them from science, I'll home-school them instead.'

Such a parent will not enjoy the Reason Rally. In 2008, at a conference of American science educators in Atlanta, Georgia, one teacher reported that students 'burst into tears' when told they would be studying evolution. Another teacher described how students repeatedly screamed 'No!' when he began talking about evolution in class.* If you are such a student, the Reason Rally is not

---

in this country have had enough of experts' and 'There's only one expert that matters, and that's you, the voter.' These examples were quoted by Michael Deacon (*Telegraph*, 10 June 2016), who went on in satirical vein: 'The mathematical establishment have done very nicely, thank you, out of the notion that $2 + 2 = 4$. Dare to suggest that $2 + 2 = 5$, and you'll be instantly shouted down. The level of group-think in the arithmetical community is really quite disturbing. The ordinary pupils of Britain, quite frankly, are tired of this kind of mathematical correctness.'

* American middle-school teachers (of ten- to fourteen-year-olds) are especially vulnerable to this kind of unpleasantness. Unlike high-school science teachers, most of them don't have science degrees and they may know too little about the overwhelming evidence in favour of evolution. Understandably they feel ill-equipped to argue, and therefore often stint the teaching of evolution or even avoid it altogether. My charitable

for you – unless you take the precaution of stopping up your ears lest a word of unwelcome truth penetrate.

'When I am faced with a mystery, with something I don't understand, I don't interrogate science for a solution, but jump to the conclusion that it must be supernatural and has no solution.'

This has been the lamentable but understandable first recourse of humanity for most of our history. We have grown out of it only during the past few centuries. Many people have never grown out of it, and if you are one of those the Reason Rally will have no appeal for you.

That is the fourth time in this essay I have said something like: 'The Reason Rally is not for you.' But let me end on a more positive note. Even if you are unaccustomed to living by reason, if you are one of those, perhaps, who actively distrust reason, why not give it a try? Cast aside the prejudices of upbringing and habit, and come along anyway. If you come with open ears and open curiosity you will learn something, will probably be entertained and may even change your mind. And that, you will find, is a liberating and refreshing experience.

A hundred years from now, there should be no need for a

---

foundation created, as one of its flagship enterprises, the Teacher Institute for Evolutionary Science (TIES). TIES exists to equip middle-school teachers with the confidence to teach evolution. It is directed by Bertha Vazquez, herself a middle-school teacher of truly outstanding qualities. She knows the problems her colleagues face and she knows her evolutionary science. At the time I write this (December 2016), she and her staff of TIES volunteers have already conducted twenty-seven workshops for middle-school teachers in states including Arkansas, North Carolina, Georgia, Texas, Florida and Oklahoma, and the numbers are increasing all the time. The participants emerge buttressed with the confidence of reliable knowledge and supplied with material teaching resources such as PowerPoint presentations prepared by Bertha and her staff.

Reason Rally. Meanwhile, unfortunately, the need is all around us and may become increasingly apparent in this election year.* Please come to Washington and stand up for reason, science and truth.

* I little realized quite how prophetic that sentence would prove to be.

In praise of subtitles; or, a drubbing for dubbing*

A LEGEND OF UNCERTAIN provenance has it that Winston Churchill, addressing a French audience about lessons learned from looking back on his own past, inadvertently raised a laugh: '*Quand je regarde mon derrière, je vois qu'il est divisé en deux parties égales.*' Most Anglos know enough French to get the joke. But alas, our knowledge doesn't go much further than Churchill's own. Whatever languages we may have learned at school – French and German in my case (as well as classical Greek and Latin, which probably influenced the way I was taught modern languages†) – we may be able to read a bit, but our spoken language performance should mantle us in shame.

---

* This essay releases a long-buzzing bee from my bonnet. Exasperation finally led me to publish it in *Prospect*, August 2016. The editors, as editors often do, shortened it somewhat. Here is the unabridged version.

† Yesterday I had lunch with an erudite classical scholar, who intrigued me by saying that, although he can read Latin and Greek exactly as fast and fluently as English, he is incapable of holding a conversation in either of those ancient languages. He can't understand spoken Latin, because the continuous stream of phonemes elides words which, on paper, are separated by spaces. He added that he has the same problem with French, and attributes it, as do I, to being taught modern languages in the same way as British schools have always taught Latin.

When I visit universities in Scandinavia or the Netherlands it goes without saying that everybody there speaks English fluently, and actually rather better than most native speakers. The same applies to almost everyone I meet outside the university: shop-keepers, waiters, taxi drivers, bartenders, random people I stop in the street to ask the way. Can you imagine a visitor to England addressing a London cabbie in French or German? And you'd have little more luck with a Fellow of the Royal Society.

The conventional explanation goes like this, and there's probably something in it. Precisely because English is so widely spoken, we don't *need* to learn any other language. Biologists like me tend to be suspicious of 'need' as an explanation for anything. A long-discredited alternative to Darwinism, Lamarckism, invoked 'need' as the driver of evolution: ancestral giraffes *needed* to reach high foliage and their energetic striving to do so somehow called longer necks into existence. But for 'need' to translate itself into action there has to be another step in the argument. The ancestral giraffe mightily stretched its neck upwards and so the bones and muscles lengthened and ... well, you know the rest, O my Best Beloved.* The true Darwinian mechanism, of course, is that those individual giraffes that succeeded in satisfying the need survived to pass on their genetic tendency to do so.

It is conceivable that a student's perceived career-need to learn

---

* My Kipling *hommage* was one of the cuts made by *Prospect*. As I explained in the essay on 'Universal Darwinism' (see pages 125–7 above), the mistaken idea of inheritance of acquired charac-teristics is a central plank of the Lamarckian theory. I've sometimes toyed with the idea of writing a Darwinian version of the *Just So Stories*, but I doubt that I (or indeed anyone other than Kipling) could pull it off. Don't be confused here by the fact that some biologists have used 'Just So Stories' as a pejorative for retrospective Darwinian rationalizations of natural phenomena. Those authors were emphasizing a different aspect of Kipling's explanations: the fact that they are retrospective. My point – that they are Lamarckian – is a separate one.

English provided the causal mechanism of a redoubled effort in the classroom. And it is possible that we whose native language is English take a deliberate decision not to bother with other languages. As a young scientist I took remedial German lessons to help me participate in international conferences, and a colleague explicitly said, 'Oh, you don't want to do that. It'll only *encourage* them.' But I doubt that most of us are that cynical.

I think the following alternative explanation should be taken seriously if only because, unlike the 'need' hypothesis, it offers the possibility of doing something about it. Again we start with the premiss that English actually is much more widely spoken than any other European language. But the next stage in the argument is different. The world is continually bombarded by English (especially American) films, songs, TV shows and soap operas. All Europeans are daily exposed to English, and they pick up English in something like the way any child learns her native language. The infant doesn't strive to satisfy a perceived 'need' to communicate. She effortlessly picks up her native language *because it is there*. Even adults can learn in something like the same way, although we lose part of our childhood ability to absorb language.* My point is that we Anglos are largely deprived of daily exposure to any language other than our own. Even when we travel abroad, we have a hard time improving our language skills because so many whom we meet are eager to speak English.

And the 'immersion' theory, unlike the 'need' theory, prompts a remedy for our monoglottish disgrace. We can change the policy of our television stations. Night after night on British TV we'll see news footage of a foreign politician, football manager, police spokesman, tennis player, or random vox pop in the street. We are allowed a few seconds of French or German, say. But then the authentic voice fades and is drowned by that of an interpreter

---

* As Steven Pinker, in *The Language Instinct*, reminds us, tiny children are linguistic geniuses at an age when they can't tie their own shoelaces.

(technically not true dubbing but 'lectoring'). I've even heard this happen when the original speaker is a great orator or statesman: General de Gaulle, say. That is lamentable for a reason over and above the main point of this article. In the case of a historic statesman, we want to hear the orator's own voice – the cadences, the emphases, the dramatic pauses, the calculated switches from strong passion to confidential quiet. And we can get these though we may not understand the words. We do *not* want the expression-less voice of a technical interpreter, or even an interpreter who makes an effort towards a more dramatic rendering. A Laurence Olivier or Richard Burton might make a better orator than a General de Gaulle, but it is the statesman we want to hear. How sincere is he? Does he mean it, or is he just playing to the gallery? How is the audience reacting to his speech? And how well is he taking their reactions? Quite apart from all that, and returning to my main point, even when the speaker is no de Gaulle but an ordinary citizen interviewed in the street, we want the opportunity to learn French, or German, or Spanish or whatever it is, in something like the way so many Europeans pick up English every day from their television news.

The power of the 'immersion effect' is incidentally demonstrated by the memetic spread of American expressions to Britain. And the 'upspeak' of British and American youth, whereby statements sound like questions, can probably be traced to the popularity of Australian soap operas. I believe it is the same process, inflated to the level of language itself, which accounts for the proficiency in English of many European nations.

When it comes to the cinema, countries are divided into those that dub and those that subtitle. Germany, Spain and Italy have dubbing cultures. It's been suggested that this is because the transition from silent films to talkies took place under dictator-ships bombastically eager to promote the national language. The Scandinavians and Dutch, by contrast, use subtitles. I'm told that German audiences recognize the voice of 'the German Sean Connery', say, as readily as we recognize Connery's own distinctive voice. True dubbing of this kind is a highly skilled and very expensive process, involving meticulous attention to lip-synched detail.*

For feature films there may be respectable defences of dubbing, although I always prefer subtitles. But I'm in any case not talking about dubbing in the expensive, lip-synching world of feature films and television drama. I'm talking about the daily ephemera of news broadcasting where the choice is between two cheap alternatives: subtitling on the one hand, or fade-out plus voiceover lectoring on the other. It is my contention that there is no decent defence of the voiceover policy. Subtitles are quite simply always better.

It's ridiculous to doubt that there's enough time to prepare subtitles for news stories. Almost all the news coverage we see is not live but rolling repeats, so there's plenty of time to write subtitles. Even for live transmissions, and even setting aside (still imperfect) computer translation, speed of preparing subtitles is not a problem. The only remotely serious argument I have ever heard in favour of the voiceover is that blind people can't read subtitles. But deaf people can't hear voiceovers, and in any case modern technology offers serviceable solutions to both handicaps. I strongly suspect that if you ask TV executives to justify their policy you'll get nothing better than, 'We've always done it and it's simply never occurred to us to use subtitles.'†

* And German film-makers do it extremely well, as I am discovering in my campaign to improve my German by watching dubbed films that I already know very well in English, such as *Jeeves und Wooster*, and *Das Leben des Brian*.

† Indeed, after I wrote that sentence I chanced to meet socially an extremely senior BBC executive and he said it in almost exactly the same words. I shamelessly made my pitch. I met him again some months later, and he told me he'd taken it to heart and was hoping to do something about it. He seemed to think it would require some kind of technical wizardry to produce subtitles quickly. I'm sceptical about that because, as I pointed out above, most TV news footage consists of rolling repeats, allowing plenty of time for subtitles to be produced by human translators.

There are those who say they 'prefer' voiceover to subtitles. My 'General de Gaulle paragraph' was, I suppose, an expression of personal preference the other way. But personal preferences vary and are often evenly balanced anyway. I want to make the case that frivolous personal preferences should be outweighed by serious educational advantages which go in only one direction. I strongly suspect that a change to a sustained subtitling policy would improve our language skills and go some way towards relieving our national shame.

AFTERWORD

A few months after this piece was published, I wrote another article for *Prospect* in which I said I was making an effort to improve my German. The reason I gave – somewhat but not wholly tongue-in-cheek – was that I was 'ashamed to be English', mostly because of the xenophobia that drove the Brexit vote, but also because of the poor language skills of my nation.

If I ruled the world . . .

HOW OFTEN DO we petulantly mutter something equivalent to: 'If I ruled the world, I'd . . .' Yet when an editor offers you the same self-indulgence out of the blue,* the mind goes blank. Frivolous answers are easy enough to reel off: ban chewing gum, baseball caps and burqas, and equip all trains with mobile-phone jammers. But such pettiness is unworthy of the editor's generosity. How about the other extreme, the utopian, pie-in-the-sky decree of universal happiness, and abolishing starvation, crime, poverty, disease and religion? Too unrealistic. So here's a manageably modest yet still worthwhile ambition: if I ruled the world, I would downgrade rulebooks and replace them, wherever possible, with humane, intelligent discretion.

I'm writing this on a plane, having just passed through security at Heathrow. A nice young mother was distraught because she wasn't allowed to take on board a tub of ointment for her little girl's eczema. The security man was polite but firm. She wasn't even permitted to spoon a reduced quantity into a smaller jar. I couldn't grasp what was wrong with that suggestion, but the rules were unbendable. The official offered to fetch his supervisor, who came and was equally polite, but she too was bound by the rulebook's hoops of steel.†

* The editors of *Prospect* magazine conceived the idea of commissioning a number of writers to muse on the title 'If I ruled the world . . .' My contribution was published in March 2011.

† I later had a similar experience myself, when I tried to take a tiny jar of what was obviously honey on a plane. Unfortunately my

There was nothing I could do, and it was no help that I recommended a website where a chemist explains, in delightfully comedic detail, what it would actually take to manufacture a workable bomb from binary liquid ingredients, labouring for several hours in the aircraft loo, using copious quantities of ice in relays of champagne coolers helpfully passed through the door by the cabin staff.

The prohibition against taking more than very small quantities of liquids or unguents on planes is demonstrably ludicrous. It started as one of those 'Look at us, we're taking decisive action' displays, the ones designed to cause maximum inconvenience to the public just to make the dimwitted dundridges* who rule our lives feel important and look busy.

It's the same with having to take our shoes off (another gem of official wally-hood that must have had Bin Laden chuckling victoriously into his beard) – and all those other exercises in belated stable-door shutting. But let me get to the general principle. Rulebooks are themselves put together by human judgements.

tweet on the subject was widely interpreted as a selfish grumble about my own precious honey, as opposed to the altruism of my concern for the young mother with the ointment. Actually, in both cases I was making a general point, an altruistic point – the very point, indeed, of this essay. I don't eat honey, by the way.

* This is a word I am grooming for – as I hope – eventual inclusion in the *Oxford English Dictionary*. I coined it from a Tom Sharpe novel, *Blott on the Landscape*, brilliantly adapted by Malcolm Bradbury for the BBC, starring Geraldine James, David Suchet and George Cole. One of the characters, 'J. Dundridge', was an epitome of the humourless, rule-following bureaucrat. In order to qualify for inclusion in the *OED*, a new coining such as '**dundridge**, *noun*' has to be used a significant number of times without definition or attribution. My footnote violates the requirement but the article in *Prospect* didn't, so I hope it counts. There's already a good word, 'jobsworth', which means the same thing, but I prefer the sound of 'dundridge'.

Often bad ones, but in any case judgements made by humans who were probably no wiser or better qualified to make them than the individuals who subsequently have to put them into practice out in the real world.

No sane person, witnessing that scene at the airport, seriously feared this woman was planning to blow herself up on a plane. The fact that she was accompanied by children gave us the first clue. Supporting evidence trickled in from the brazen visibility of her face and hair, from her lack of a Quran, prayer mat or big black beard and, finally, from the absurdity of the notion that her tub of ointment could, in a million years, be magicked into a high explosive – certainly not in the cramped facilities afforded by an aircraft loo. The security official and his supervisor were human beings who obviously wished they could behave decently, but they were powerless: stymied by a rulebook. Nothing but an object, which, because it is made of paper and unalterable ink rather than of flexible human brain tissue, is incapable of discretion, compassion or humanity.

This is just a single example and it may seem trivial. But I am sure that you, dear reader, can list half a dozen similar cases from your own experience.* Talk to any doctor or nurse, and hear their frustration with having to spend a substantial proportion of their

---

* An eight-year-old boy of my acquaintance begged his parents to let him join them in running a 10-kilometre race. They demurred, agreeing with the rulebook that he was too young. However, he was so disappointed that they agreed to let him start, assuming that he would honourably drop out early during the race and one of them would drop out with him. In the event, he didn't drop out but kept up with his father all the way, and beat his mother – no slouch herself. But when he reached the finishing line, the officials wouldn't let him cross it. He was under the age limit, and was made to go round the side. Maybe they should have pulled him out on the starting line. But to do that to a child in his moment of triumph as he reached the finishing line was – well – stable doors and bolting horses come to mind again.

time filling in forms and ticking boxes. Who sincerely thinks that is a good use of expert, valuable time; time that could be spent caring for patients? No human being, surely – not even a lawyer. Only a mindless book of rules.

How often does a criminal walk free on a 'technicality'? Perhaps the arresting officer fluffed his lines when delivering the official caution. Decisions that will gravely affect a person's life can turn on the powerlessness of a judge to exercise discretion and reach the conclusion that every single person in the court, often even including the accused and his defence lawyer, knows is just.

It isn't as simple as this, of course. Discretion can be abused, and rulebooks are important safeguards against that. But the balance has shifted too far in the direction of an obsessive reverence for rules. There must be ways to reintroduce intelligent discretion and overthrow the unbending tyranny of going by the book without opening the door to abuse. If I ruled the world, I would make it my business to find them.*

---

* Another example of officials playing by the rulebook, although a moment of reflection would have shown them how ridiculous it was, is the story concerning my uncle Colyear Dawkins and the Oxford station barrier: see the Afterword to my eulogy to my Uncle Bill on page 412.

VI

THE SACRED TRUTH
OF NATURE

THE TITLE OF THIS SECTION echoes a comment in the volume's opening essay: for scientists, 'there is something almost sacred about nature's truth'. There, the context was the sanctity of truth in science; here, I use the phrase to herald a group of pieces celebrating that truth expressed in actuality, through observation of the glories and complexities of the natural world. At its core are two essays that take their texts from that richest of ecological hotspots, the ultimate place of pilgrimage for the ardent Darwinian: the Galápagos Islands.

We begin, though, not on an equatorial beach but in high abstraction, with the concept of time, as addressed in a lecture opening an exhibition 'About time'. This and the closing piece of the section share a lyrical, even elegiac reflectiveness, punctuated by an affectionate delight in the quirks and oddities of the natural world, the ridiculously fascinating and fascinatingly ridiculous – such as the Pacific Palolo worm, which engages in communal self-amputation for breeding purposes, or the amnesiacally flightless kakapo, launching itself in ambitious panic from a tree and collapsing in a heap on the floor.

The theme of time continues into the two 'tales' that follow – their titles recalling the constituent parts of that most inventive and encyclopedic of Richard's works, *The Ancestor's Tale*. These were written on a trip to the islands in 2005, and are infused with the pilgrim's delight in this richly surreal Arcadia. Around the eponymous central figure of each – the giant tortoise, the turtle – are woven discussions of life's twisting journey from water to land

(and sometimes back again) over the unimaginable expanses of geological time.

The section concludes with a foreword to a wonderful book celebrating this fragile paradise and the fragility of the wider world's biodiversity: a revised edition of *Last Chance to See* by Douglas Adams and Mark Carwardine. If this essay has a melancholic tone about it, that is hardly surprising. Not only is the book for which it was written itself an elegy for vanishing species driven to the edge of extinction, but at the time of its composition the writer, along with countless others, was still mourning the untimely death of Douglas Adams – humorist, humanist, celebrator of science – at the tragically young age of forty-nine. This is both a paean to the priceless riches of our living planet and a threnody for a priceless human being.

G.S.

About time*

TIME IS PRETTY mysterious stuff – almost as elusive and hard to pin down as conscious awareness itself. It seems to flow – 'like an ever-rolling stream' – but what is it that does the flowing? We have the feeling that the present is the only instant of time that actually exists. The past is a shadowy memory, the future a vague uncertainty. Physicists don't see it like that. The present has no privileged status in their equations. Some modern physicists have gone so far as to describe the present as an illusion, a product of the observer's mind.

For poets, time is anything but an illusion. They hear its wingèd chariot hurrying near; they aspire to leave footprints on the sands of it; wish there was more of it – to stand and stare; invite it to put up its caravan, just for one day. Proverbs declare procrastination to be the thief of it; or they compute, with improbable precision, the ratio of stitches saved in it. Archeologists excavate rose-red cities half as old as it. Pub landlords announce it gentlemen please. We waste it, spend it, eke it out, squander it, kill it.

Long before there were clocks or calendars, we – indeed all

* The Ashmolean Museum is Oxford's leading museum of art and antiquities. In 2001 it staged an exhibition called 'About Time', featuring clocks and timekeepers through the ages. I felt honoured by being invited to open it, and this is the speech with which I did so. The text was subsequently published in the *Oxford Magazine*, 2001.

animals and plants – measured out our lives by the cycles of astronomy. By the wheeling of those great clocks in the sky: the rotation of the Earth on its axis, the rotation of the Earth around the sun, and the rotation of the moon around the Earth.

By the way, it's surprising how many people think the Earth is *closer* to the sun in summer than in winter. If this were really so, Australians would have their winter at the same time as ours. A glaring example of such Northern Hemisphere chauvinism was the science-fiction story in which a group of space travellers, far out in some distant star system, waxed nostalgic for the home planet: 'Just to think that it's spring back on Earth!'

The third great clock in our sky, the orbiting of the moon, exerts its effects on living creatures mostly via the tides. Many sea creatures order their lives according to a lunar calendar. The Pacific Palolo worm, *Palolo viridis* or *Eunice viridis*, lives in crevices of coral reefs. In the early mornings of two particular days during the last quarter of the moon in October, the rear ends of all the worms simultaneously break off and swim to the surface for a breeding frenzy. These are remarkable rear ends. They even have their own pair of eyes.

The same thing happens twenty-eight days later, in the last quarter of the November moon. So predictable is the timing that the islanders know exactly when to go out in their canoes and gather up the squirming rear ends of Palolo worms, which are a prized delicacy.

Notice that the Palolo worms achieve their synchrony not by simultaneously responding to a particular signal from the sky. Rather, each worm independently *integrates* cycles registered over many lunar cycles. They all do the same sums on the same data, so like good scientists they all come to the same conclusion and break off their rear ends simultaneously.

A similar story could be told of plants synchronizing their flowering seasons by integrating successively measured changes in day length. Many birds time their breeding seasons in the same way. This is easily demonstrated by experiments using artificial lights put on and off by time-switches to simulate artificial day-lengths appropriate to different times of the year.

Most animals and plants – probably all living cells – have internal clocks buried deep in their biochemistry. These biological clocks manifest themselves in all kinds of physiological and behavioural rhythms. You can measure them in dozens of different ways. They are linked to the external astronomical clocks, and normally synchronized to them. But the interesting thing is that if the biological clocks are separated from the outside world, they carry on regardless. They truly are *internal* clocks. Jet-lag is the discomfort we experience when our own internal clocks are being reset by the external *Zeitgeber*** after a major shift of longitude.

Longitude is, of course, intimately linked with time. John Harrison's winning solution to the great longitude competition of the eighteenth century was nothing more than a clock that stayed accurate even when taken to sea. Migrating birds, too, make use of their own internal clocks for similar navigational purposes.

Here's a lovely example of an internal clock. As you know, worker bees have a code with which they tell fellow hive members where they have found food. The code is a figure-of-eight dance, which they perform on the vertical comb inside the hive. There is a straight run in the middle of the figure of eight, whose direction conveys the direction of the food. Since the dance is performed on the vertical comb, whereas the angle of the food is in the horizontal plane, there has to be a convention. The convention is that the upward direction on the comb in the vertical plane stands for the sun's direction in the horizontal plane. A dance with a straight run straight up the comb tells the other bees to leave the hive and fly dead towards the sun. A dance with the straight run 30° to the right of the vertical on the comb tells the other bees: Leave the hive and fly at an angle 30° to the right of the sun.

* The use of the German word for time-giver or synchronizer in the scientific literature reflects the fact that much of the classic work in the field was done in Germany.

Well, that is remarkable enough, and when Karl von Frisch first discovered it, many people found it hard to believe. But it is true.* And it gets even better, and this brings us back to the sense of time. There's a problem with using the sun as a reference point. It moves. Or rather, since the Earth spins, the sun appears to move (from left to right in the Northern Hemisphere), as the day advances. How do the bees cope?

Von Frisch tried the experiment of trapping his bees in his observation hive for several hours. They went on dancing. But he noticed something which really is almost too good to be true. As the hours advanced, the dancing bees slowly turned the direction

---

* How it arose in evolution is a fascinating question. Von Frisch and his colleagues have compared the dance to various more primitive equivalents in other species of bee. Some nest out in the open and signal the direction of food by repeating a 'take-off run' in the horizontal plane, pointing directly towards the food source that they have discovered. Think of it as a kind of 'follow me in this direction' gesture, repeated several times to recruit more followers. But how did this get translated into the code used on the vertical comb, where 'up' (against gravity) in the vertical plane stands for 'direction of the sun' in the horizontal plane? There's a clue in an odd quirk of the insect nervous system, demonstrated in insects as distantly related to each other as beetles and ants. First, a piece of background information (not the quirk): as I mentioned on page 259, many insects use the sun as a compass, flying in a straight line by keeping the sun at a fixed angle. This is easily demonstrated using an electric light to simulate the sun. Now for the quirk. Experimenters watched their insect as it walked over a horizontal surface, maintaining a fixed angle to an artificial light source. They then switched off the light, simultaneously tilting the horizontal surface into the vertical. The insect continued to walk, but switched its direction so that the angle to the vertical was the same as the previous angle to the light. I call it a quirk because the circumstance is unlikely to arise in nature. It is as though there is some kind of wire-crossing in the insect nervous system, which was convenient for exploitation in the evolution of the bee dance.

of the straight run of their dance, so that it would continue to tell the truth about the direction of the food, compensating for the *changing* position of the sun. And they did this, even though they were dancing inside the hive and therefore couldn't see the sun. They were using their internal clocks to compensate for what they 'knew' would be the changing position of the sun.

What this means, if you think about it, is that the straight run of the dance itself rotates like the hour hand of a normal clock (though at half the speed). But anticlockwise (in the Northern Hemisphere), like the shadow on a sundial. If you were von Frisch, wouldn't you have died happy, to have made such a discovery?

Even after clocks were invented, sundials remained essential for setting clocks and keeping them synchronized with the great clock in the sky. Hilaire Belloc's famous rhyme is, therefore, rather unfair:

> I am a sundial, and I make a botch
> Of what is done far better by a watch.

It is less well-known that Belloc wrote a whole series of verses on sundials, some humorous, some sombre, more in keeping with the 'Fighting Time' theme of our exhibition:

> How slow the Shadow creeps: but when 'tis past
> How fast the Shadows fall. How fast! How fast!

> Creep, shadow, creep: my ageing hours tell.
> I cannot stop you, so you may as well.

> Stealthy the silent hours advance, and still;
> And each may wound you, and the last shall kill.

> Save on the rare occasions when the Sun
> Is shining, I am only here for fun.

> I am a sundial, turned the wrong way round.
> I cost my foolish mistress fifty pound.

You may think of this last verse when you look round the exhibition and see the exquisite little pocket sundial. It has a built-in compass, without which it would be useless.

When I talked of the great clocks in the sky, I did not go out beyond one year, but there are potential astronomical clocks of hugely longer period. Our sun takes about two hundred million years to complete one rotation around the centre of the galaxy. As far as I am aware, no biological process has become entrained to this cosmic clock.*

The longest timekeeper that has been seriously suggested as being influential on life is an approximately twenty-six-million-year periodicity of mass extinctions. The evidence for this involves sophisticated statistical analysis of extinction rates in the fossil record. It is controversial and by no means definitely demonstrated. There is no doubt that mass extinctions happen, and at least one of them is pretty likely to have been caused by the impact of a comet, sixty-five million years ago when the dinosaurs perished. More controversial is the idea that such events rise to a peak of likelihood every twenty-six million years.†

Another suggested astronomical clock longer than a year is the eleven-year sunspot cycle, which might account for certain cycles in populations of Arctic mammals, such as lynxes and snowshoe hares, as detected by Charles Elton, that great Oxford ecologist, in fur-trapping records of the Hudson's Bay Company. This theory, too, remains controversial.

Director, you invited a biologist to perform this opening, so you will not be surprised to have been regaled with stories about bees and Palolo worms and snowshoe hares. You could have asked

---

* Indeed, I would be very surprised if one were ever found.

† My speech made mention of a hypothetical astronomical clock to account for it, but I have deleted it from this reprinting because modern astronomers mostly discount it and there is no direct evidence for it. Briefly, the suggestion was that the sun mutually orbits a binary companion star, called Nemesis, with a periodicity of about twenty-six million years. The gravitational effect of Nemesis was supposed to disturb the Oort cloud of planetesimals and increase the probability of one hitting Earth.

an archeologist, and we'd all have been engrossed in tales of dendrochronology, or of radiocarbon dating. Or a paleontologist, and we'd have heard about potassium–argon dating, and about the near-impossibility, for the human mind, of grasping the sheer vastness of geological time. The geologist would have used one of those metaphors with which we struggle – and usually fail – to understand geological deep time. My own favourite one I didn't invent, I hasten to add, although I did use it in one of my books. As follows:

> Fling your arms wide to represent the whole history of evolution from the origin of life at your left fingertip to the present day at your right fingertip. All the way across your midline to well past your right shoulder, life consists of nothing but bacteria. Animal life begins to flower somewhere around your right elbow. The dinosaurs originate in the middle of your right palm, and go extinct around your last finger joint. The whole story of *Homo sapiens* and our predecessor *Homo erectus* is contained in the thickness of one nail-clipping. As for recorded history; as for Babylon, as for the Assyrian who came down like a wolf on the fold, as for the Jewish patriarchs, the legions of Rome, the Christian Fathers, the dynasties of Pharaohs, the Laws of the Medes and Persians which never change; as for Troy and the Greeks; as for Napoleon and Hitler, the Beatles and the Spice Girls, they and everyone that knew them are blown away in the dust from one light stroke of a nail-file.

If I had been a historian, I would have told stories of how different peoples have perceived time. Of how some cultures see it as cyclical, others as linear, and how this influences their whole attitude to life. Of how the Islamic calendar is based upon the lunar cycle, where ours is annual. Of how clocks used to be made, in the days before Galileo used his own heart as a clock to work out the Law of the Pendulum, and engineers perfected escapements. I would have added that the Chinese had an escapement clock, driven by water, as early as the tenth century AD.

I would have remarked how the calibration of Egyptian water clocks had to be different at different times of year, because the

Egyptian hour was defined as one twelfth part of the time between dawn and dusk – so one summer hour was longer than one winter hour. Richard Gregory, from whom I learned this singular fact, remarks, mildly, that 'this must have given the Egyptians a rather different sense of time from ours . . .'

If I had been a physicist or cosmologist, my reflections on time would have been perhaps most remarkable of all. I would have tried – and probably failed – to explain that the Big Bang was not only the beginning of the universe, but the beginning of time itself. To the obvious question, what happened before the Big Bang, the answer – or so physicists try in vain to persuade us – is that it is simply an illegitimate question. The word 'before' can no more be applied to the Big Bang than you can walk north from the North Pole.

If I had been a physicist, I would have tried to explain that, in a vehicle travelling at an appreciable fraction of the speed of light, time itself slows down – as perceived from outside the vehicle, though not within it. If you travelled through space at such prodigious speeds you could return to Earth five hundred years into the future, having yourself scarcely aged at all. This is not some therapeutic effect of high-speed travel upon the human constitution. It is an effect upon time itself. Contrary to Newtonian cosmology, time is not absolute.

Some physicists are even prepared to contemplate true time travel, going backwards in time – which I suppose must be any historian's dream. I find it almost comical that one of the main arguments against this is the element of paradox. Suppose you killed your own great-grandmother!* Science-fiction writers have responded by giving their time travellers a rigid code of conduct. Every time traveller must swear an oath not to mess about with

---

* You could do something far less drastic to change the course of history such that you would never be born. A sneeze would do it, given the prior improbability that any particular one out of billions of spermatozoa would succeed in fertilizing an egg.

history. Somehow one feels that nature herself must erect stronger barriers than fickle human laws and conventions.

If I had been a physicist, I would also have considered the symmetry or asymmetry of time. How deep is the distinction between a process running forwards in time and one running backwards? How fundamental is the difference between a film running backwards or forwards? The laws of thermodynamics seem to provide an asymmetry. Famously, you can't unscramble an egg; and a shattered glass does not spontaneously reassemble itself.

Does biological evolution reverse the thermodynamic arrow? No, for the law of increasing entropy applies only to closed systems, and life is an open system, driven upstream by energy from outside. But evolutionists, too, have their own version of the question whether time has an arrow of direction. Is evolution progressive?

Well, I may not be a physicist but I am an evolutionary biologist, and you had better not get me started on *that* fascinating question.

One of the things that any speaker can do with time is run out of it. The important business of the evening is to look at this exhibition 'About Time'. I was privileged to be shown around it yesterday, and I can tell you it is fascinating – in all sorts of ways. It gives me very great pleasure to declare the exhibition open.

AFTERWORD

Reading this speech again, I recognize how tantalizingly brief my scientific vignettes on time might have seemed – not long enough to explain anything properly. My excuse is that it was my job to tantalize: and to encourage guests to proceed to the exhibition and think about time while enjoying it.

There are those who say, incidentally, that the Ashmolean should be called the Tradescantian Museum because it was originally founded to house the collections, mostly of natural history, made by the John Tradescants, father and son. The Tradescant collections were acquired (some say through dubious means) by Elias Ashmole (1617–92), who bequeathed them to

Oxford University, which continued to add to them. The Tradescant natural history collections were transferred in the 1850s to the newly built University Museum of Natural History, and the Ashmolean became mostly an art museum.

There is also an argument – a different one – for changing the name of the Natural History Museum too, because so many visitors to Oxford think its name is 'Pitt Rivers'. Though annexed to the main museum building, the Pitt Rivers Museum is an entirely separate institution with a remarkable collection of anthropological artefacts, grouped not by region, as is customary, but by function: fishing nets all together, flutes all together, timekeepers all together, and so on. To avoid the popular confusion with the Pitt Rivers, I have suggested renaming the Museum of Natural History the Huxley Museum. 'Tradescantian' would redress a seventeenth-century injustice but would open up a new confusion. The Huxley Museum would commemorate the alleged 'victory' of T. H. Huxley over Bishop Sam Wilberforce in the 'Great Debate' which took place in the newly built museum building. I must say I have mixed feelings about that because there is reason to think the scale of the 'victory' has been exaggerated.

The giant tortoise's tale: islands within islands*

I AM WRITING THIS on a boat in the Galápagos archipelago, whose most famous inhabitants are the eponymous giant tortoises, and whose most famous visitor is that giant of the mind, Charles Darwin. In his account of the voyage of HMS *Beagle*, written long before the central idea of *The Origin of Species* focused itself in his brain, Darwin wrote of the Galápagos islands:

> Most of the organic productions are aboriginal creations, found nowhere else; there is even a difference between the inhabitants of the different islands; yet all show a marked relationship with those of [South] America, though separated from that continent by an open space of ocean, between 500 and 600 miles in width. The archipelago is a little world within itself . . . Considering the small size of the islands, we feel the more astonished at the number of their aboriginal beings, and at their confined range . . . we seem to

---

* *The Ancestor's Tale*, now in its second edition jointly authored with Yan Wong, was first published shortly before I went on a memorable trip to the Galápagos Islands as the grateful guest of Victoria Getty. The book's central motif is a 'pilgrimage' to the past. The *hommage* to Chaucer extends to 'Tales' told by particular animals, each tale conveying a general biological message. The momentum to tell such 'tales' persisted into my visit to the Galápagos, whose fauna moved me while on board ship to write three extra tales which were published in the *Guardian*. This one appeared on 19 February 2005.

be brought somewhat near to that great fact – that mystery of mysteries – the first appearance of new beings on this earth.

True to his pre-Darwinian education, the young Darwin was using 'aboriginal creation' for what we would now call endemic species – evolved on the islands and found nowhere else. Nevertheless, Darwin already had more than a faint inkling of that great truth with which, in his mighty maturity, he was to enlighten the world. Writing of the small birds now known as Darwin's Finches, he said:

> Seeing this gradation and diversity of structure in one small, intimately related group of birds, one might really fancy that from an original paucity of birds in this archipelago, one species had been taken and modified for different ends.

He could as well have said the same of the giant tortoises, for he was told by the Vice-Governor, Mr Lawson, that

> the tortoises differed from the different islands, and that he himself could with certainty tell from which island any one was brought. I did not for some time pay sufficient attention to this statement, and I had already partially mingled together the collections from two of the islands. I never dreamed that islands, about 50 or 60 miles apart, and most of them in sight of each other, formed of precisely the same rocks, placed under a quite similar climate, rising to a nearly equal height, would have been differently tenanted.

And he said the same kind of thing about the iguanas, both marine and land, and the plants.

With the benefit of hindsight, *Darwinian* hindsight, we post-Darwinians can piece together what happened. In every one of these cases – and this is typical of the origin of species everywhere – it is islands that constitute the vital – though accidental – ingredient. Without the isolation provided by islands, sexual intermingling of gene pools nips species divergence in the bud. Any aspiring new species would be continually flooded by genes from the old species. Islands are natural workshops of evolution. A barrier to sexual intermingling is what you need, to allow that initial divergence of gene pools which constitutes the origin of species, Darwin's 'Mystery of Mysteries'.

But islands don't have to be land surrounded by water. The tortoise's tale has two lessons to teach us, and this is the first one. To a highland-breeding giant tortoise, each of the five volcanoes along the length of the big island of Isabela (Albemarle to Darwin, who used the traditional English names) is an island of green habitability surrounded by inhospitable lava desert. Most of the Galápagos Islands are a single volcano each, so the two kinds of islands coincide. But the big island, Isabela, is a necklace of five volcanoes, spaced from each other at approximately the same distance as the single volcano on the neighbouring island of Fernandina which, from one point of view, might as well be a sixth volcano on Isabela. To a tortoise, Isabela is an archipelago within an archipelago.

Both levels of isolation have played a role in the evolution of the giant tortoises. All the Galápagos giant tortoises are related to a particular mainland species of land tortoise, *Geochelone chilensis*, which still survives and is smaller than any of them. At some point during the few million years that the islands have existed, one or a few of these mainland tortoises inadvertently fell in the sea and floated across. How could it have survived the long and doubtless arduous crossing without food or fresh water? Well, the early whalers took thousands of giant tortoises from the Galápagos Islands to their ships for food. To keep the meat fresh, the tortoises were not killed until needed. But they were not fed or watered while waiting to be butchered. They were simply turned on their backs so they couldn't walk away. I tell the story not in order to horrify (although I have to say that it horrifies me), but to make a point. Tortoises can survive for weeks without food or fresh water, easily long enough to float in the Humboldt Current from South America to the Galápagos Islands. And tortoises do float.

Having reached the archipelago, the tortoises did what many animals do when they arrive on an island. They evolved to become

larger: the long noticed phenomenon of island gigantism.* If the tortoise story had followed the finch pattern, they would have evolved a different species on each of the islands. Then, if there were subsequent accidental driftings from island to island, they would have been unable to interbreed (that's the definition of a separate species) and would have been free to evolve a different way of life from their colleagues of different species on the new island, and also from their colleagues of the same species on other islands.†
In the case of the finches, you could say that the different species' incompatible mating habits and preferences now constitute a kind of genetic substitute for the geographic isolation of separate islands. Though they overlap geographically, they are isolated on separate islands of mating exclusivity. So they can diverge yet further. Most of the Galápagos Islands have the large, the medium and the small ground finch, which specialize in different diets. These three species surely originally diverged on different islands and have now come together where they coexist as different species on the same islands, never interbreeding and each specializing in a different kind of seed diet.

The tortoises did something similar,‡ evolving distinctive shell shapes on the different islands. The tortoises on the larger islands

---

* Confusingly, island dwarfism is also common. There were dwarf elephants on several Mediterranean islands and dwarf hominins, *Homo floresiensis*, on the small Indonesian island of Flores.

† There are also giant tortoises on the island of Aldabra in the Indian Ocean. And there were others, until nineteenth-century sailors drove them extinct along with the dodo and its cousins, on Mauritius and neighbouring islands. The Indian Ocean tortoises show the same evolutionary phenomenon of island gigantism as the Galápagos ones, but they evolved independently, in their case from smaller ancestors that drifted from Madagascar.

‡ But without the second stage of coming together again to share the same island, after diverging.

tend to have high domes. Those on the smaller islands have saddle-shaped shells with a high-lipped aperture for the head at the front. The reason for this seems to be that the large islands usually have enough water to grow grass, and the tortoises there are grazers. On the smaller islands there is often not enough water to grow grass, and the tortoises have to become browsers on cactuses. The high-lipped saddle shell allows the neck to reach up to the cactuses. The cactuses, for their part, grow taller and taller in an evolutionary arms race against the browsing tortoises.

The tortoise story adds to the finch model the further complication we have already noted: for them, volcanoes are islands within islands. They provide high, cool, damp, green oases, surrounded by dry lava fields at low altitude which, for a giant tortoise, constitute hostile desert. Most of the islands have but a single volcano and each has its own single species (or sub-species) of giant tortoise (some have none at all). The big island of Isabela has five major volcanoes, and each of them has its own species (or sub-species) of tortoise. Truly, Isabela is an archipelago within an archipelago. And the principle of archipelagoes as powerhouses of divergent evolution has never been more elegantly demonstrated than here in the islands of Darwin's blest youth.

The sea turtle's tale: there and back again (and again?)*

IN 'THE GIANT tortoise's tale' I described ancestral tortoises float-ing inadvertently across from South America, colonizing the Galápagos Islands by mistake, subsequently evolving local differ-ences on each island and giant size on all of them. But why assume that the colonizer was a land tortoise? Wouldn't it be simpler to guess that marine turtles, already at home in the sea, hauled up on the island beaches as if to lay their eggs, enjoyed what they saw, stayed on dry land and evolved into tortoises? No. Nothing like that happened on the Galápagos Islands, which have only been in existence a few million years.

Something very like it did happen, however, much longer ago in the ancestry of all tortoises. But that anticipates the climax to the turtle's tale. (By the way, the word 'turtle' is a tiresome example of Bernard Shaw's observation that England and America are two countries divided by a common language. In British usage, turtles live in water and tortoises on land. For Americans, tortoises are those turtles that live on land.)

There is good evidence that the most recent common ancestor of all today's land tortoises, including those on the mainlands of America, Australia, Africa and Eurasia, as well as the giants of Galápagos, Aldabra, the Seychelles and other oceanic islands, was itself a land tortoise. In their less ancient ancestry, to misquote

---

* This was the second of my additional tales, written on board ship in Galápagos and published in the *Guardian*, 26 February 2005.

Stephen Hawking, it's tortoises all the way down. The various giant tortoises of the Galápagos Islands are certainly descended from South American land tortoises.

If you go back far enough everything lived in the sea: watery alma mater of all life. At various points in evolutionary history, enterprising individuals within many different animal groups moved out onto the land, sometimes even to the most parched deserts, taking their own private sea water with them in blood and cellular fluids. In addition to the reptiles, birds, mammals and insects which we see all around us, other groups that have succeeded out of water include scorpions, snails, crustaceans such as woodlice and land crabs, millipedes and centipedes, spiders and their kin, and various worms. And we mustn't forget the plants, without whose prior invasion of the land none of the other migrations could have happened.

This was an immense journey to undertake, not necessarily in terms of geographic distance but in terms of the upheaval in every aspect of life, from breathing to reproduction. Among the vertebrates, a particular group of lobe-finned fishes, related to today's coelacanths and lungfishes, took to walking on land and developed lungs for breathing air. Their descendants the reptiles developed a large egg with a waterproof shell to retain the moisture that, from ancestral times in the sea, all vertebrate embryos need. Later descendants of the early reptiles included mammals and birds, which evolved a wide range of techniques for exploiting the land environment, including the habit of living in deserts, revolutionizing their way of life so that it became about as different from the ancestral life in the sea as can be imagined.

Among the wide range of specializations displayed by land creatures was one that seems wilfully perverse: a good number of thoroughgoing land animals later turned around, abandoned their hard-earned terrestrial re-tooling, and trooped back into the water again. Seals and sea lions (such as the astonishingly tame Galápagos sea lion) have only gone part-way back. They show us what the intermediates might have been like, on the way to extreme cases such as whales and dugongs. Whales (including the small whales we call dolphins), and dugongs with their close cousins the manatees, ceased to be land creatures altogether and reverted to the full marine habits of their remote ancestors. They don't even come ashore to breed. They do, however, still breathe air, having never developed anything equivalent to the gills of their earlier marine incarnation.

Other animals that have returned from land to water are pond snails, water spiders, water beetles, Galápagos flightless cormorants, penguins (Galápagos has the only penguins in the Northern Hemisphere*), marine iguanas (found nowhere but Galápagos) and turtles (abundant in the surrounding waters).

Iguanas are adept at surviving accidental oceanic crossings on driftwood (well documented within the West Indies), and there can be no doubt that the marine iguanas of Galápagos trace back to just such a piece of living flotsam from South America. The oldest of the existing Galápagos Islands is no older than about four million years. Since the marine iguanas evolved here and nowhere else, you might think this sets a maximum limit on the date of their return to the water. The story is more complicated, however.

The Galápagos Islands were made, one after the other, as the Nazca tectonic plate moved, at a rate of a few centimetres per year, over a particular volcanic hotspot under the Pacific Ocean. As the plate moved east, from time to time the hotspot punched through, delivering another island along the production line. This is why the youngest islands are towards the west and the oldest to the east. But, at the same time as the Nazca plate continues to move east, it is also being subducted under the South American plate. The easternmost islands sink under the sea, at a rate of about one centimetre per

* On my most recent visit to Galápagos, our senior Ecuadorian guide told an amusing story. A previous guest on the ship had raved about the experience – the scenery, the natural history, the food, the boat. He had only one complaint: Galápagos penguins are too small.

year. It is now known that, although the oldest existing island is only four million years old, there has been an eastward-moving and sinking archipelago in this area for at least seventeen million years. Islands now submerged could have provided the initial haven for iguanas to colonize and evolve, at any time during that period. There would have been plenty of time for them to island-hop before their original ancestral island sank beneath the waves.

Turtles went back to the sea much longer ago. They are, in one respect, less fully given back to the water than whales or dugongs, for turtles still lay their eggs on beaches. Like all vertebrate returners to the water, they breathe air, but in this department they go one better than whales. Some turtles extract additional oxygen from the water through a pair of chambers at the rear end, richly supplied with blood vessels. One Australian river turtle, indeed, gets the majority of its oxygen by breathing, as an Australian would not hesitate to say, through its arse.

There is evidence that all modern turtles are descended from a terrestrial ancestor who lived before most of the dinosaurs. There are two key fossils called *Proganochelys quenstedti* and *Palaeochersis talampayensis* dating from early dinosaur times, which appear to be close to the ancestry of all modern turtles and tortoises. You might wonder how we tell whether fossil animals, especially if only fragments are found, lived on land or in water. Sometimes it's pretty obvious. Ichthyosaurs were reptilian contemporaries of the dinosaurs, with fins and streamlined bodies. The fossils look like dolphins and they surely lived like dolphins, in the water. With turtles it is a little less obvious. One neat way to tell is by measuring the bones of their forelimbs.

Walter Joyce and Jacques Gauthier, at Yale University, took three key measurements in the arm and hand bones of seventy-one species of living turtles and tortoises. They used triangular graph paper to plot the three measurements against one another. Lo and behold, all the land tortoise species formed a tight cluster of points in the upper part of the triangle; all the water turtles clustered in the lower part of the triangular graph. There was no overlap, except when they added some species that spend time in both water and land. Sure enough, these amphibious species show up, on the triangular graph, halfway between the 'wet cluster' and the 'dry cluster'. Well then, to the obvious next step: where do the fossils fall? The hands of *P. quenstedti* and *P. talampayensis* leave us in no doubt. Their points on the graph are right in the thick of the dry cluster. Both these fossils were dry-land tortoises. They come from the era before our turtles returned to the water.

You might think, therefore, that modern land tortoises have probably stayed on land ever since those early terrestrial times, as most mammals did after a few of them went back to sea. But apparently not. If you draw out the family tree of all modern turtles and tortoises, nearly all the branches are aquatic. Today's land tortoises constitute a single branch, deeply nested among branches consisting of aquatic turtles. This suggests that modern land tortoises have not stayed on land continuously since the time of *P. quenstedti* and *P. talampayensis*. Rather, their ancestors were among those who went back to the water, and they then re-emerged back onto the land in (relatively) more recent times.

Tortoises therefore represent a remarkable double return. In common with all mammals, reptiles and birds, their remote ancestors were marine fish and before that various more or less worm-like creatures stretching back, still in the sea, to the primeval bacteria. Later ancestors lived on land and stayed there for a very large number of generations. Even later ancestors evolved back into water-dwellers and became sea turtles. And finally they returned yet again to the land as tortoises, some of which, though not the Galápagos giants, now live in the driest of deserts.

I have described DNA as 'the genetic book of the dead' (see also page 82). Because of the way natural selection works, there is a sense in which the DNA of an animal is a textual description of the worlds in which its ancestors were naturally selected. For a fish, the genetic book of the dead describes ancestral seas. For us humans and most other mammals, the early chapters of the book are all set in the sea and the later ones all out on land. For whales, dugongs, marine iguanas, penguins, seals, sea lions, turtles and, remarkably, tortoises, there is a third section of the book which recounts their epic return to the proving grounds of their remote past, the sea. But for the tortoises, perhaps uniquely, there is yet a fourth section of the book devoted to a final – or is it? – re-emergence, yet again to the land. Can there be another animal for whom the genetic book of the dead is such a palimpsest of multiple evolutionary U-turns?

Farewell to a digerati dreamer

M<small>Y OWN LAST</small> chance to see* Douglas Adams in action as a public speaker was at the Digital Biota conference in Cambridge in September 1998. As it happens, I dreamed last night of a similar event: a small conference of like-minded people, Douglas's kind of people, denizens of the wild 'Here be Digerati' badlands between zoology and computer technology, one of Douglas's favourite habitats. He was there of course, holding court (as I saw it, although his large and generously jocular modesty would have mocked the phrase). I had that familiar dreaming sense of knowing that he was dead, but of finding it not the least bit odd that he should be among us anyway, talking about science and making us laugh with his uniquely scientific wit. He was eagerly telling us over lunch about a remarkable adaptation in a fish, and he informed us that it would need only twenty-seven mutations to evolve it from a trout. I wish I could remember what the remarkable adaptation was, for it was exactly the kind of thing Douglas would have read about somewhere, and 'twenty-seven mutations' is exactly the kind of detail he would have relished.

From Cambridge to Komodo (from digerati to dragons) is no big step for a dreamer, so perhaps Douglas's fish was the mudskipper that prompted his ancestral reflections at the end of the Komodo

---

* A play on words, for this essay was first published as the foreword to the 2009 edition of Douglas Adams and Mark Carwardine, *Last Chance to See*.

dragon chapter. His use of mudskippers and their 350-million-year-old forerunners – and ours – to tie up the dragon chapter and assuage his nagging guilt at not having spoken up for the hapless goat is a literary *tour de force*. Even the unfortunate chicken comes back as a metaphor, to reprise its tragi-comic role as the uneasy starter before the main course of pathetically bleating goat.

> It is an uncomfortable experience to share a long ride on a small boat with four live chickens who are eyeing you with a deep and dreadful suspicion which you are in no position to allay.

Nobody's written like that since P. G. Wodehouse. Or like this:

> a benign man with the air of a vicar apologizing for something.

Or like this, on a rhinoceros grazing:

> It was like watching a JCB excavator quietly getting on with a little weeding ... The animal measured about six feet high at its shoulders, and sloped down gradually towards its hindquarters and its rear legs, which were chubby with muscle. The sheer immensity of every part of it exercised a fearful magnetism on the mind. When the rhino moved a leg, just slightly, huge muscles moved easily under its heavy skin like Volkswagens parking ... The rhino snapped to attention, turned away from us, and hurtled off across the plain like a nimble young tank.

That last phrase is pure PGW, but Douglas had the advantage of an additional, scientific dimension to his humour. Wodehouse could never have achieved this:

> It felt as if we were participating in a problem of three-body physics, swinging round in the gravitational pull of the rhinos.

Or this, of the Philippines monkey-eating eagle:

> a wildly improbable-looking piece of flying hardware that you would more readily expect to see coming in to land on an aircraft carrier than nesting in a tree.

Chapter One's reverie on 'Twig Technology' is original enough to provoke a scientist to serious thought, as is Douglas's meditation on the rhinoceros as an animal whose world is dominated by smell, instead of vision. Douglas was not just knowledgeable about science. He didn't just make jokes about science. He had the mind of a scientist, he mined science deeply and brought to the surface . . . humour, and a style of wit that was simultaneously literary and scientific, and uniquely his own.

There is probably no page in this book that doesn't set me laughing out loud whenever I reread it – which is even more often than I read his fiction. In addition to the witty language, there are wonderful passages of sustained set-piece comedy, as in the epic quest for a condom in Shanghai (to sheathe an underwater microphone for listening to Yangtze River dolphins). Or there's the legless taxi driver who kept diving under the dashboard to operate the clutch with his hands. Or there's the wry comedy of the bureaucrats of Mobuto's Zaire, whose corrupt nastiness exposes in Douglas and his comrade Mark Carwardine a benign innocence that recalls the kakapo, out of its depth in a harsh and uncaring world:

> The kakapo is a bird out of time. If you look one in its large, round, greeny-brown face, it has a look of serenely innocent incomprehension that makes you want to hug it and tell it that everything will be all right, though you know that it probably will not be.
>
> It is an extremely fat bird. A good-sized adult will weigh about six or seven pounds, and its wings are just about good for waggling a bit if it thinks it's about to trip over something. Sadly, however, it seems that not only has the kakapo forgotten how to fly, but it has also forgotten that it has forgotten how to fly. Apparently a seriously worried kakapo will sometimes run up a tree and jump out of it, whereupon it flies like a brick and lands in a graceless heap on the ground.

The kakapo is one of several island species of animals that, in the interpretation offered here, are ill-equipped to hold their own against predators and competitors whose gene pools have been honed in the harsher ecological climate of the mainland:

> So you can imagine what happens when a mainland species gets introduced to an island. It would be like introducing Al Capone,

Genghis Khan and Rupert Murdoch into the Isle of Wight – the locals wouldn't stand a chance.

Of the endangered animals that Douglas Adams and Mark Carwardine set out to see, one seems to have gone for good during the intervening two decades. We have now lost our last chance to see the Yangtze River dolphin. Or hear it, which is more to the point, for the river dolphin lived in a world where seeing was pretty much out of the question anyway: a murky, muddy river in which sonar came splendidly into its own – until the arrival of massive noise pollution by the engines of boats.

The loss of the river dolphin is a tragedy, and some of the other wonderful characters in this book cannot be far behind. In his Last Word, Mark Carwardine reflects on why we should care when species, or whole major groups of animals and plants, go extinct. He deals with the usual arguments:

> Every animal and plant is an integral part of its environment: even Komodo dragons have a major role to play in maintaining the ecological stability of their delicate island homes. If they disappear, so could many other species. And conservation is very much in tune with our own survival. Animals and plants provide us with life-saving drugs and food, they pollinate crops and provide important ingredients for many industrial processes.

Yes, yes, we have to say that kind of thing, it's expected of us. But the pity is that we *need* to justify conservation on such human-centred, utilitarian grounds. To borrow an analogy I've used in a different context, it's a bit like justifying music on the grounds that it's good exercise for the violinist's right arm. Surely the real justification for saving these magnificent creatures is the one with which Mark rounds off the book, and which he obviously prefers:

> There is one last reason for caring, and I believe that no other is necessary. It is certainly the reason why so many people have devoted their lives to protecting the likes of rhinos, parakeets, kakapos and dolphins. And it is simply this: the world would be a poorer, darker, lonelier place without them.

Yes!

The world is a poorer, darker, lonelier place without Douglas Adams. We still have his books, his recorded voice, memories, funny stories, affectionate anecdotes. I literally cannot think of another departed public figure whose memory arouses such universal affection, among those who knew him personally and those who didn't. He was especially loved by scientists. He understood them and was able to articulate, far better than they could, what gets their blood running. I used that very phrase, in a television documentary called *Break the Science Barrier*, when I interviewed Douglas and asked him: 'What is it about science that really gets your blood running?' His impromptu reply should be framed on the wall of every science classroom in the land:

> The world is a thing of utter inordinate complexity and richness and strangeness that is absolutely awesome. I mean the idea that such complexity can arise not only out of such simplicity, but probably absolutely out of nothing, is the most fabulous extraordinary idea. And once you get some kind of inkling of how that might have happened – it's just wonderful. And ... the opportunity to spend seventy or eighty years of your life in such a universe is time well spent as far as I am concerned.*

Seventy or eighty? If only.

The pages of this book sparkle with science, scientific *wit*, science seen through the rainbow prism of 'a world-class imagination'. There is no cloying sentimentality in Douglas's view of the aye-aye, the kakapo, the northern white rhino, the echo parakeet, the Komodo dragon. Douglas understood very well how slowly

---

* The documentary was for Channel 4 and was broadcast in 1996. In the course of the interview, Douglas had just said that in the nineteenth century the novel was where one would go for 'serious reflections about life', but that nowadays 'the scientists actually tell us much much more about such issues than you'd ever get from novelists'. I then asked him: 'What is it about science that really gets your blood running?' and this was his reply.

grind the mills of natural selection. He knew how many megayears it takes to build a mountain gorilla, a Mauritius pink pigeon or a Yangtze River dolphin. He saw with his own eyes how quickly such painstaking edifices of evolutionary artifice can be torn down and tossed to oblivion. He tried to do something about it. So should we, if only to honour the memory of this unrepeatable specimen of *Homo sapiens*. For once, the specific name is well deserved.

VII

LAUGHING AT LIVE DRAGONS

IN A WAY IT'S A FALSE categorization to devote a specific section of this book to humour. If you've read consecutively thus far, you'll know why: even on the gravest of topics, where it issues in a hue pretty much as dark as black humour gets, and irrepressibly in lighter contexts, humour is a constantly glinting seam through the Dawkins oeuvre. So why this section? It's always been a puzzle to me, and indeed something of an irritation, to read this or that interview or profile and find the writer saying something along the lines that 'Richard Dawkins is of course a very clever man but has no sense of humour' or 'the trouble with atheists is, they have no sense of humour'. This is so blatantly wrong that it seems justifiable – and in harmony with the scientific method – to offer a little evidence.

Exhibits A–G here, chosen to reflect Richard Dawkins' own heroes of comic writing as well as his own considerable talent in its practice, range from pitch-perfect pastiche to prodigal inventiveness to the pithiest of ironies. All have in common the wit and linguistic agility that run through so much of the material in this book; here that seam of gold hits the surface.

It was the search for gold, of course, that woke the dragon in Tolkien's fantasy story *The Hobbit*; and it was the courageous 'everyman' Bilbo who cautioned himself: 'Never laugh at live dragons.' Richard would have no truck with fear of fire-breathing monsters; but his eagerness to prod the ferocious as well as the ridiculous might well raise a wizard's eyebrow.

Pastiche and satire alike require a sensitive ear for voice as well as a practised hand with language; pastiche *as* satire requires a particularly sure touch, and 'Fundraising for faith' is so close to the voice of the keen New Labour acolyte as to make it hard not to feel for the blushes of the bright young things of the sometime PM's office who surely must recognize their argot coming back at them.

Crafted with an equally sure touch, and wearing lightly their weighty messages – debunking the theology of the Atonement and outlining the mechanism of evolution by natural selection – the two Wodehouse parodies, 'The Great Bus Mystery' and 'Jarvis and the Family Tree', are sheer delight in their homage to a master of Englishness, right down to the aunt's footstep on the stair.

Satire, of course, can be deadly serious as well as wincingly funny, as witnessed here most forcefully by the next piece, 'Gerin Oil'. In view of Richard's dedication to the frequently thankless task of bearing reason's banner into hostile territory, it's surely an achievement of some order to retain not only a lively sense of irony but a lightness of touch on even the grimmest of subjects.

There is plenty of companionable humour here, too – laughing with the dragon-hunters and indeed the dragon-lovers. From P. G. Wodehouse to Robert Mash, the 'Sage elder statesman of the dinosaur fancy', there's a heritage here of literate wordiness, a fellowship of lovers of language and what it can do, in which Richard is undeniably at home. In his foreword to Mash's *How to Keep Dinosaurs* he sets out his own allegiances in literary humour and then, with evident delight and enjoyment, enters into the parallel world, picks up the baton and adds his own flourishing coda.

Finally, after the rich diet of dinosaur comes the astringent brevity of two crisp satires. 'Athorism: let's hope it's a lasting vogue' turns the language and argumentation of modern theology back on itself with evident glee and consummate skill; and to round off the section, 'Dawkins' Laws' clothe wry frustration in the garb of philosophical discourse and nail an important truth with pinpoint wit.

Even Gandalf might be impressed.

G.S.

Fundraising for faith*

DEAR PERSON OF FAITH

Basically, I write as fundraiser for the wonderful new Tony Blair Foundation, whose aim is 'to promote respect and understanding about the world's major religions and show how faith is a powerful force for good in the modern world'. I would like to touch base with you on six key points from the recent *New Statesman* piece by Tony (as he likes to be called by everybody, of all faiths – or indeed of none, for that's how tuned in he is!).

* Tony Blair sank from extreme popularity to the reverse, purely on the strength of his devotion to George W. Bush and their disastrous war in Iraq. History will be kinder to the pair of them, if only by comparison with what we are about to experience in 2017 and for the next four years. I've even heard American friends gloomily enunciating, 'Come back, Bush, all is forgiven.' And Tony Blair is re-emerging as a voice of sanity in Brexit-blighted Blighty. However, Blair's immediate activity, on leaving office, was to found a preposterous charity for the promotion of religious faith. It didn't seem to matter which faith you supported. Faith itself was supposed to be a good thing, to be encouraged. I published this satire on his foundation, complete with the style of English that has become known as mediaspeak, in the *New Statesman*, 2 April 2009. It's a point-by-point spoof response to an article Blair himself had written in the same journal.

'My faith has always been an important part of my politics.'

Yes indeed, although Tony modestly kept shtum about it when he was PM. As he said, to shout his faith from the rooftops might have been interpreted as claiming moral superiority over those with no faith (and therefore no morals, of course). Also, some might have objected to their PM taking advice from voices only he could hear; but hey, reality is so last year compared with private revelation, no? What else, other than shared faith, could have brought Tony together with his friend and comrade-in-arms, George 'Mission Accomplished' Bush, in their life-saving and humanitarian intervention in Iraq?

Admittedly, there are one or two problems remaining to be ironed out there, but all the more reason for people of different faiths – Christian and Muslim, Sunni and Shia – to join together in meaningful dialogue to seek common ground, just as Catholics and Protestants have done, so heart-warmingly, throughout European history. It is these great benefits of faith that the Tony Blair Foundation seeks to promote.

'We are focusing on five main projects initially, working with partners in the six main faiths.'

Yes I know, I know, it's a pity we had to limit ourselves to six. But we do have boundless respect for other faiths, all of which, in their colourful variety, enrich human lives.

In a very real sense, we have much to learn from Zoroastrianism and Jainism. And from Mormonism, though Cherie says we need to go easy on the polygamy and the sacred underpants!! Then again, we mustn't forget the ancient and rich Olympian and Norse traditions – although our modern blue-skies thinking out of the box has pushed the envelope on shock-and-awe tactics, and put Zeus's thunderbolts and Thor's hammer in the shade!!! We hope, in Phase 2 of our Five-Year Plan, to embrace Scientology and Druidic Mistletoe Worship, which, in a very real sense, have something to teach us all. In Phase 3, our firm commitment to Diversity will lead us to source new networking partnership opportunities with the many hundreds of African tribal religions. Sacrificing goats may present problems with the RSPCA, but we hope to persuade them to adjust their priorities to take proper account of religious sensibilities.

'We are working across religious divides towards a common goal – ending the scandal of deaths from malaria.'

Plus, of course, we mustn't forget the countless deaths from AIDS. This is where we can learn from the Pope's inspiring vision, expounded recently on his visit to Africa. Drawing on his reserves of scientific and medical knowledge – informed and deepened by the *Values* that only faith can bring – His Holiness explained that the scourge of AIDS is made worse, not better, by condoms. His advocacy of abstinence may have dismayed some medical experts (and the same goes for his deeply and sincerely held opposition to stem-cell research). But surely to goodness we must find room for a diverse range of opinions. All opinions, after all, are equally valid, and there are many ways of knowing, spiritual as well as factual. That, at the end of the day, is what the Foundation is all about.

'We have established Face to Faith, an interfaith schools programme to counter intolerance and extremism.'

The great thing is to foster diversity, as Tony himself said in 2002, when challenged by a (rather intolerant!!!!) MP about a school in Gateshead teaching children that the world is only six thousand years old. Of course you may think, as Tony himself happens to, that the true age of the world is 4.6 billion years. But – excuse me – in this multicultural world, we must find room to tolerate – and indeed actively foster – all opinions: the more diverse, the better. We are looking to set up video-conferencing dialogues to brainstorm our differences. By the way, that Gateshead school ticked lots of boxes when it came to GCSE results, which just goes to show.

'Children of one faith and culture will have the chance to interact with children of another, getting a real sense of each other's lived experience.'

Cool! And, thanks to Tony's policy of segregating as many children as possible in faith schools where they can't befriend kids from other backgrounds, the need for this interaction and mutual understanding has never been so strong. You see how it all hangs together? Sheer genius!

So strongly do we support the principle that children should be sent to schools which will identify them with their parents' beliefs, that we think there is a real opportunity here to broaden it out. In Phase 2, we look to facilitate separate schools for Postmodernist children, Leavisite children and Saussurian Structuralist children. And in Phase 3 we shall roll out yet more separate schools, for Keynesian children, Monetarist children and even neo-Marxist children.

'We are working with the Coexist Foundation and Cambridge University to develop the concept of Abraham House.'

I always think it's so important to coexist, don't you agree, with our brothers and sisters of the other Abrahamic faiths. Of course we have our differences – I mean, who doesn't, basically? But we must all learn mutual respect. For example, we need to understand and sympathize with the deep hurt and offence that a man can feel if we insult his traditional beliefs by trying to stop him beating his wife, or setting fire to his daughter or cutting off her clitoris (and please don't let's hear any racist or Islamophobic objections to these important expressions of faith). We shall support the introduction of Sharia courts, but on a strictly voluntary basis – only for those whose husbands and fathers freely choose it.

'The Blair Foundation will work to leverage mutual respect and understanding between seemingly incompatible faith traditions.'

After all, despite our differences, we do have one important thing in common: all of us in the faith communities hold firm beliefs in the total absence of evidence, which leaves us free to believe anything we like. So, at the very least, we can be united in claiming a privileged role for all these private beliefs in the formulation of public policy.

I hope this letter will have shown you some of the reasons why you might consider supporting Tony's Foundation. Because hey, let's face it, a world without religion doesn't have a prayer. With so many of the world's problems caused by religion, what better solution could there possibly be than to promote yet more of it?

The Great Bus Mystery*

I WAS HOOFING IT down Regent Street, admiring the Christmas decorations, when I saw the bus. One of those bendy buses that mayors keep threatening with the old heave-ho. As it drove by, I looked up and got the message square in the monocle. You could have knocked me down with a feather. Another of the blighters nearly did knock me down as I set a course for the Dregs Club, where it was my purpose to inhale a festive snifter, and I saw the same thing on the side. There are some pretty deep thinkers to be found at the Dregs, as my regular readers know, but none of them could make a dent on the vexed question of the buses when I

* In 2009 the journalist and comedian Ariane Sherine initiated a campaign to promote atheism on British buses. My foundation (RDFRS UK) helped to fund it, along with the British Humanist Association, and we were involved in planning it. The wording of the slogan on the buses was Ariane's own, and I think it was excellent: 'There's probably no God. Now stop worrying and enjoy your life.' The word 'probably' came in for some criticism, but I think it worked perfectly: intriguing enough to provoke discussion, while disavowing unjustified confidence. At the end of the same year, Ariane edited a lovely Christmas anthology called *The Atheist's Guide to Christmas*. For my contribution, I paid tribute to her bus campaign in the form of a parody of my favourite comic author. For copyright reasons, I was advised by m'learned friend to disguise the names of the characters.

bowled it their way. Not even Swotty Postlethwaite, the club's tame intellectual. So I decided to put my trust in a higher power.

'Jarvis,' I sang out, as I latchkeyed self into the old headquarters, shedding hat and stick on my way through the hall to consult the oracle. 'I say, Jarvis, what about these buses?'

'Sir?'

'You know, Jarvis, the buses, the "What is this that roareth thus?" * brigade, the bendy buses, the conveyances with the kink amidships. What's going on? What price the bendy bus campaign?'

'Well, sir, I understand that, while flexibility is often considered a virtue, these particular omnibuses have not given uniform satisfaction. Mayor Johnson . . .'

'Never mind Mayor Johnson, Jarvis. Consign Boris to the back burner and bend the bean to the buses. I'm not referring to their bendiness *per se*, if that is the right expression.'

'Perfectly correct, sir. The Latin phrase might be literally construed . . .'

'That'll do for the Latin phrase. Never mind their bendiness. Fix the attention on the slogan on the side. The orange and pink apparition that flashes by before you have a chance to read it properly. Something like "There's no bally God, so put a sock in it and have a gargle with the lads." That was the gist of it, anyway, although I may have foozled the fine print.'

'Oh yes, sir, I am familiar with the admonition: "There's probably no God. Now stop worrying and enjoy your life." '

'That's the baby, Jarvis. Probably no God. What's it all about? Isn't there a God?'

---

* The first line of a once famous comic verse by A. D. Godley, full of Latin rhyming jokes designed to appeal to Englishmen of Bertie's class who would have learned Latin at school: http://latindiscussion.com/forum/latin/a-d-godleys-motor-bus.10228/. 'Bendy buses' was the nickname given to the articulated buses introduced into London in the early 2000s and later controversially withdrawn from service by Mayor Boris Johnson.

'Well, sir, some would say it depends upon what you mean. All things which follow from the absolute nature of any attribute of God must always exist and be infinite, or, in other words, are eternal and infinite through the said attribute. Spinoza.'

'Thank you, Jarvis, I don't mind if I do. Not one I've heard of, but anything from your shaker always hits the s. and reaches the parts other cocktails can't. I'll have a large Spinoza, shaken not stirred.'

'No, sir, my allusion was to the philosopher Spinoza, the father of pantheism, although some prefer to speak of panentheism.'

'Oh, *that* Spinoza, yes, I remember he was a friend of yours. Seen much of him lately?'

'No, sir, I was not present in the seventeenth century. Spinoza was a great favourite of Einstein, sir.'

'Einstein? You mean the one with the hair and no socks?'

'Yes, sir, arguably the greatest physicist of all time.'

'Well, you can't do better than that. Did Einstein believe in God?'

'Not in the conventional sense of a personal God, sir, he was most emphatic on the point. Einstein believed in Spinoza's God, who reveals himself in the orderly harmony of what exists, not in a God who concerns himself with fates and actions of human beings.'

'Gosh, Jarvis, bit of a googly* there, but I think I get your drift. God's just another word for the great outdoors, so we're wasting our time lobbing prayers and worship in his general direction, what?'

'Precisely, sir.'

'If, indeed, he has a general direction,' I added moodily, for I can spot a deep paradox as well as the next man, ask anyone at the Dregs. 'But Jarvis,' I resumed, struck by a disturbing thought. 'Does this mean I was also wasting my time when I won that prize for

* Cricket, of course. A googly is a spun ball where the bowler's hand action misleads the batsman as to the direction of spin. Devious spin bowlers sometimes intersperse googlies in among other, more conventionally spun balls.

Scripture Knowledge at school? The one and only time I elicited so much as a murmur of praise from that prince of stinkers, the Rev. Aubrey Upcock? The high spot of my academic career, and it turns out to have been a dud, a washout, scratched at the starting gate?'

'Not entirely, sir. Parts of holy writ have great poetic merit, especially in the English translation known as the King James, or Authorized version of 1611. The cadences of the Book of Ecclesiastes and some of the prophets have seldom been surpassed, sir.'

'You've said a mouthful there, Jarvis. Vanity of vanities, saith the Preacher. Who was the preacher, by the way?'

'That is not known, sir, but informed opinion agrees that he was wise. Rejoice, O young man, in thy youth; and let thy heart cheer thee in the days of thy youth. He also evinced a haunting melancholy, sir. When the grasshopper shall be a burden, and desire shall fail: because man goeth to his long home, and the mourners go about the streets. The New Testament too, sir, is not without its admirers. For God so loved the world that he gave his only begotten Son . . .'

'Funny you should mention that, Jarvis. The passage was the very one I raised with the Rev. Aubrey, and it provoked a goodish bit of throat-clearing and shuffling of the trotters.'

'Indeed, sir. What was the precise nature of the late headmaster's discomfort?'

'All that stuff about dying for our sins, redemption and atonement. All that "and with his stripes we are healed" carry-on. Being, in a modest way, no stranger to stripes administered by old Upcock, I put it to him straight. "When I've performed some misdemeanour" – or malfeasance, Jarvis?'

'Either might be preferred, sir, depending on the gravity of the offence.'

'So, as I was saying, when I was caught perpetrating some malfeasance or misdemeanour, I expected the swift retribution to land fairly and squarely on the Woofter trouser seat, not some other poor sap's innocent derrière, if you get my meaning?'

'Certainly, sir. The principle of the scapegoat has always been of dubious ethical and jurisprudential validity. Modern penal theory casts doubt on the very idea of retribution, even where it is the malefactor himself who is punished. It is correspondingly harder to justify vicarious punishment of an innocent substitute. I am pleased to hear that you received proper chastisement, sir.'

'Quite.'

'I am so sorry, sir, I did not intend . . .'

'Enough, Jarvis. This is not dudgeon. Umbrage has not been taken. We Woofters know when to move swiftly on. There's more. I hadn't finished my train of thought. Where was I?'

'Your disquisition had just touched upon the injustice of vicarious punishment, sir.'

'Yes, Jarvis, you put it very well. Injustice is right. Injustice hits the coconut with a crack that resounds around the shires. And it gets worse. Now, follow me like a puma here. Jesus was God, am I right?'

'According to the Trinitarian doctrine promulgated by the early Church Fathers, sir, Jesus was the second person of the Triune God.'

'Just as I thought. So God – the same God who made the world and was kitted out with enough nous to dive in and leave Einstein gasping at the shallow end, God the all-powerful and all-knowing creator of everything that opens and shuts, this paragon above the collarbone, this fount of wisdom and power – couldn't think of a better way to forgive our sins than to turn himself over to the gendarmerie and have himself served up on toast. Jarvis, answer me this. If God wanted to forgive us, why didn't he just forgive us? Why the torture? Whence the whips and scorpions, the nails and the agony? Why not just forgive us? Try that on your Victrola.'

'Really, sir, you surpass yourself. That is most eloquently put. And if I might take the liberty, sir, you could even have gone further. According to many highly esteemed passages of traditional theological writing, the primary sin for which Jesus was atoning was the Original Sin of Adam.'

'Dash it, you're right. I remember making the point with some vim and *élan*. In fact, I rather think that may have been what tipped the scales in my favour and handed me the jackpot in that Scripture Knowledge fixture. But do go on, Jarvis, you interest me strangely. What was Adam's sin? Something pretty fruity, I imagine. Something calculated to shake hell's foundations?'

'Tradition has it that he was apprehended eating an apple, sir.'

'Scrumping?* That was it? That was the sin that Jesus had to redeem – or atone according to choice? I've heard of an eye for an eye and a tooth for a tooth, but a crucifixion for a scrumping? Jarvis, you've been at the cooking sherry. You are not serious, of course?'

'Genesis does not specify the precise species of the purloined comestible, sir, but tradition has long held it to have been an apple. The point is academic, however, since modern science tells us that Adam did not in fact exist, and therefore was presumably in no position to sin.'

'Jarvis, this takes the chocolate digestive, not to say the mottled oyster. It was bad enough that Jesus was tortured to atone for the sins of lots of other fellows. It got worse when you told me it was only one other fellow. It got worse still when that one fellow's sin turned out to be nothing worse than half-inching a D'Arcy Spice. And now you tell me the blighter never existed in the first place. I am not known for my size in hats, but even I can see that this is completely doolally.'

'I would not have ventured to use the epithet myself, sir, but there is much in what you say. Perhaps in mitigation I should mention that modern theologians regard the story of Adam, and his sin, as symbolic rather than literal.'

'Symbolic, Jarvis? Symbolic? But the whips weren't symbolic. The nails in the cross weren't symbolic. If, when I was bending over that chair in the Rev. Aubrey's study, I had protested that my misdemeanour, or malfeasance if you prefer, had been merely symbolic, what do you think he would have said?'

---

* A rather specialized verb which, I suspect, is unknown in American English, scrumping means stealing apples, raiding an orchard.

'I can readily imagine that a pedagogue of his experience would have treated such a defensive plea with a generous measure of scepticism, sir.'

'Indeed. You are right. Upcock was a tough bimbo. I can still feel the twinges in damp weather. But perhaps I didn't quite skewer the point, or nub, in re the symbolism?'

'Well, sir, some might consider you a trifle hasty in your judgement. A theologian would probably aver that Adam's symbolic sin was not so very negligible, since what it symbolized was all the sins of mankind, including those yet to be committed.'

'Jarvis, this is pure apple sauce. "Yet to be committed"? Let me ask you to cast your mind back yet again to that doom-laden scene in the beak's study. Suppose I had said, from my vantage point doubled up over the armchair, "Headmaster, when you have administered the statutory six of the juiciest, may I respectfully request another six in consideration of all the other misdemeanours, or peccadilloes, which I may or may not decide to commit at any time into the indefinite future. Oh, and make that all future misdemeanours committed not just by me but by any of my pals." It doesn't add up. It doesn't float the boat or ring the bell.'

'I hope you will not take it as a liberty, sir, if I say that I am inclined to agree with you. And now, if you will excuse me, sir, I would like to resume decorating the room with holly and mistletoe, in preparation for the annual yuletide festivities.'

'Decorate if you insist, Jarvis, but I must say I hardly see the point any more. I expect the next thing you'll tell me is that Jesus wasn't really born in Bethlehem, and there never was a stable or shepherds or wise men following a star in the East.'

'Oh no, sir, informed scholars from the nineteenth century onwards have dismissed those as legends, often invented to fulfil Old Testament prophecies. Charming legends but without historical verisimilitude.'

'I feared as much. Well, come on, out with it. Do you believe in God?'

'No, sir. Oh, I should have mentioned it before, sir, but Mrs Gregstead telephoned.'

I paled beneath the t. 'Aunt Augusta? She isn't coming here?'

'She did intimate some such intention, sir. I gathered that she proposes to prevail upon you to accompany her to church on Christmas Day. She took the view that it might improve you, although she expressed a doubt that anything could. I rather fancy that is her footstep on the stairs now. If I might make the suggestion, sir . . .'

'Anything, Jarvis, and be quick about it.'

'I have unlocked the fire escape door in readiness, sir.'

'Jarvis, you were wrong. There is a God.'

'Thank you very much, sir. I endeavour to give satisfaction.'

Jarvis and the Family Tree*

'I SAY, JARVIS, CLUSTER round.'

'Sir?'

'Close on me – if that's the right expression?'

'A military phrase, sir, employed by officers requiring the presence of their subordinates.'

'Right, Jarvis. Lend me your ears.'

'Equally appropriate, sir. Mark Antony . . .'

'Never mind Mark Antony. This is important.'

'Very good, sir.'

'As you know, Jarvis, when it comes to regions north of the collar stud, B. Woofter is not rated highly in the form book. Nevertheless, I do have one great scholastic triumph to my credit. And I bet you don't know what that was?'

'You have frequently adverted to it, sir. You won the prize for Scripture Knowledge at your preparatory academy.'

'Yes, I did, to the ill-concealed surprise of the Rev. Aubrey Upcock, proprietor and chief screw at that infamous hell-hole. And ever since then, although not much of a lad for Matins or Evensong, I've always had a soft spot for Holy Writ as we experts call it. And now we come to the nub. Or crux?'

'Very appropriate, sir, or "nitty gritty" is these days often heard.'

'The point is, Jarvis, as an *aficionado*, I have long been especially

---

* I enjoyed writing the previous parody enough to have another go the following Christmas. This one is previously unpublished.

fond of the book of Genesis. God made the world in six days, am I right?'

'Well, sir . . .'

'Beginning with light, God moved swiftly through the gears, making plants and things that creep, scaly things with fins, our feathered friends tootling through the trees, furry brothers and sisters in the undergrowth and finally, rounding into the straight, he created chaps like us, before taking to his hammock for a well-earned siesta on the seventh day. Am I right?'

'Yes, sir, if I may say so, a colourfully mixed summary of one of our great origin myths.'

'But now, Jarvis, mark the sequel. A fellow at the Dregs Club Christmas party was bending my ear last night over the snort that refreshes. Seems there's a cove called Darwin who says Genesis is all a lot of rot. God's been oversold on the campus. He didn't make everything after all. There's something called *evaluation* . . .'

'Evolution, sir. The theory advanced by Charles Darwin in his great book of 1859, *On the Origin of Species.*'

'That's the baby. Evolution. Would you credit it, this Darwin bozo wants me to believe my great-great-grandfather was some kind of hirsute banana-stuffer, scratching himself with his toes and swinging through the treetops. Now, Jarvis, answer me this. If we're descended from chimpanzees, why are there chimpanzees still among those present and correct? I saw one only last month at the zoo. Why haven't they all turned into members of the Dregs Club (or the Athenaeum according to taste)? Try that on your pianola.'

'If I might take the liberty, sir, you appear to be labouring under a misunderstanding. Mr Darwin does not say that we are descended from chimpanzees. Chimpanzees and we are descended from a shared ancestor. Chimpanzees are modern apes, which have been evolving since the time of the shared ancestor, just as we have.'

'Hm, well, I think I get your drift. Just as my pestilential cousin Thomas and I are both descended from the same grandfather. But neither of us looks any more like the old reprobate than the other, and neither of us has his side-whiskers.'

'Precisely, sir.'

'But hang on, Jarvis. We old lags of the Scripture Knowledge handicap don't give up that easily. My old man's guvnor may have been a hairy old gargoyle, but he wasn't what you'd call a chimpanzee. I distinctly remember. Far from dragging his knuckles over the ground, he carried himself with an upright, military bearing (at least until his later years, and when the port had gone round a few times). And the family portraits in the old ancestral home, Jarvis. We Woofters did our bit at Agincourt, and there were no apes on the strength during that "God for Harry, England and St George" carry-on.'

'I think, sir, you underestimate the timespans involved. Only a few centuries have passed since Agincourt. Our shared ancestor with chimpanzees lived more than five million years ago. If I might venture upon a flight of fancy, sir?'

'Certainly you might. Venture away, with the young master's blessing.'

'Suppose you walk back in time one mile, sir, to reach the Battle of Agincourt . . .'

'Sort of like walking from here to the Dregs?'

'Yes, sir. On the same scale, to walk back to the ancestor we share with chimpanzees, you'd have to walk all the way from London to Australia.'

'Goodness, Jarvis, all the way to the land of cobbers with corks dangling from their lids. No wonder there are no apes among the family portraits, no low-browed chest-thumpers to be seen once-more-unto-the-breaching at Agincourt.'

'Indeed, sir, and to go back to our shared ancestor with fish . . .'

'Wait a minute, hold it there. Are you now telling me I'm descended from something that would feel at home on a slab?'

'We share ancestors with modern fish, sir, which would certainly have been called fish if we could see them. You could safely say that we are descended from fish, sir.'

'Jarvis, sometimes you go too far. Although, when I think of Gussie Hake-Wortle . . .'

'I would not have ventured to make the comparison myself, sir. But if I might pursue my fanciful perambulation back through time, sir? To reach the ancestor that we share with our piscine cousins . . .'

'Let me guess, you'd have to walk right round the whole bally globe and come back to where you started and surprise yourself from behind?'

'A considerable underestimate, sir. You'd have to walk to the moon and back, and then set off and do the whole journey again, sir.'

'Jarvis, this is too much to spring on a lad with a morning head. Go and mix me one of those pick-me-ups of yours before I can take any more.'

'I have one in readiness, sir, prepared when I perceived the lateness of the hour when you returned from your club last night.'

'Attaboy, Jarvis. But wait, here's another thing. This Darwin bird says it all happened by chance. Like spinning the big wheel at Le Touquet. Or like when Bufty Snodgrass scored a hole in one and stood drinks for the whole club for a week.'

'No, sir, that is incorrect. Natural selection is not a matter of chance. Mutation is a chance process. Natural selection is not.'

'Take a run-up and bowl that one by me again, Jarvis, if you wouldn't mind. And this time make it your slower ball, with no spin. What is mutation?'

'I beg your pardon, sir, I presumed too much. From the Latin *mutatio*, feminine, "a change", a mutation is a mistake in the copying of a gene.'

'Like a misprint in a book?'

'Yes, sir, and, like a misprint in a book, a mutation is not likely to lead to improvement. Just occasionally, however, it does, and then it is more likely to survive and be passed on in consequence. That would be natural selection. Mutation, sir, is random in that it has no bias towards improvement. Selection, by contrast, is automatically biased towards improvement, where improvement means ability to survive. One could almost coin a phrase, sir, and say: "Mutation proposes, selection disposes."'

'Rather neat that, Jarvis. Your own?'

'No, sir, the pleasantry is an anonymous parody of Thomas à Kempis.'

'So, Jarvis, let me see if I've got a firm grip on the trouser seat of this problem. We see something that looks like a piece of natty design, like an eye or a heart, and we wonder how it bally well got here.'

'Yes, sir.'

'It can't have got here by pure chance because that would be like Bufty's hole in one, when we had drinks all round for a week.'

'In some respects it would be even more improbable than the Honourable Mr Snodgrass's alcoholically celebrated feat with the driver, sir. For all the parts of a human body to come together by sheer chance would be about as improbable as a hole in one if Mr Snodgrass were blindfolded and spun around, so that he had no idea of the whereabouts of the ball on the tee, nor of the direction of the green. Were he to be permitted a single stroke with a wood, sir, his chance of scoring a hole in one would be about as great as the chance of a human body spontaneously coming together if all its parts were shuffled at random.'

'What if Bufty had had a few drinks beforehand, Jarvis? Which, by the way, is pretty likely.'

'The contingency of a hole in one is sufficiently remote, sir, and the calculation sufficiently approximate, that we may neglect the possible effects of alcoholic stimulants. The angle subtended at the tee by the hole . . .'

'That'll do, Jarvis, remember I have a headache. What I clearly see through the fog is that random chance is a non-starter, a washout, scratched at the off. So how *do* we get complex things that work, like human bodies?'

'To answer that question, sir, was Mr Darwin's great achievement. Evolution happens gradually and over a very long time. Each generation is imperceptibly different from the previous one, and the degree of improbability required in any one generation is not prohibitive. But after a sufficiently large number of millions of generations, the end product can be very improbable indeed, and can look very much as though it was designed by a master engineer.'

'But it only *looks* like the work of some slide-rule toting whizz with a drawing board and a row of biros in his top pocket?'

'Yes, sir, the illusion of design results from the accumulation of a large number of small improvements in the same direction, each one small enough to result from a single mutation, but the whole cumulative sequence being sufficiently prolonged to culminate in an end result that could not have come about in a single chance event. The metaphor has been advanced of a slow climb up the gentle slopes of what has somewhat over-dramatically been called "Mount Improbable", sir.'

'Jarvis, that's a doosra* of an idea, and I think I'm beginning to get my eye in for it. But I wasn't too far wrong, was I, when I called it "evaluation" instead of evolution?'

'No, sir. The process somewhat resembles the breeding of racehorses. The fastest horses are *evaluated* by breeders and the best ones are chosen as progenitors of future generations. Mr Darwin realized that in nature the same principle works without the need for any breeder to do the evaluating. The individuals that run fastest are automatically less likely to be caught by lions.'

'Or tigers, Jarvis. Tigers are very fast. Inky Brahmapur was telling me at the Dregs only last week.'

'Yes, sir, tigers too. I can well imagine that his Highness would have had ample opportunity to observe their speed from the back of his elephant. The nub, or crux, is that the fastest individual horses survive to breed and pass on the genes that made them fast, because they are less likely to be eaten by large predators.'

'By Jove, that makes a lot of sense. And I suppose the fastest tigers also get to breed because they are the first ones to grab their

---

* Cricket again: another kind of deceptively spun ball, invented by the Pakistani bowler Saqlain Mushtaq. These are arcane matters and I confess to being vague about the details of how it differs from a googly.

medium rare with all the trimmings, and so survive to have little tigers that also grow up to be fast.'

'Yes, sir.'

'But this is amazing, Jarvis. This really prangs the triple twenty. And the same thing works not just for horses and tigers but for everything else?'

'Precisely, sir.'

'But wait a moment. I can see that this bowls Genesis middle stump. But where does it leave God? It sounds from what this Darwin bimbo says that there's not a lot left for God to do. I mean to say, I know what it's like to be underemployed, and underemployed is what God, if you get my drift, would seem to be.'

'Very true, sir.'

'So, well, dash it, I mean to say, in that case why do we even believe in God at all?'

'Why indeed, sir?'

'Jarvis, this is astounding. Incredulous.'

'Incredible, sir.'

'Yes, incredible. I shall see the world through new eyes, no longer through a glass darkly as we biblical scholars say. Don't bother with that pick-me-up. I find I no longer need it. I feel sort of *liberated*. Instead, bring me my hat, my stick, and the binoculars Aunt Daphne gave me last Goodwood. I'm going out into the park to admire the trees, the butterflies, the birds and the squirrels, and marvel at everything you have told me. You don't mind if I do a spot of marvelling at everything you've told me, Jarvis?'

'No indeed, sir. Marvelling is very much in the proper vein, and other gentlemen have told me that they experience the same sense of liberation on first comprehending such matters. If I might make a further suggestion, sir?'

'Suggest away, Jarvis, suggest away, we are always ready to hear suggestions from you.'

'Well, sir, if you would care to follow the matter further, I have a small volume here, which you might care to peruse.'

'Doesn't look very small to me, but anyway, what is it called?'

'It is called *The Greatest Show on Earth*, sir, and it is by . . .'

'It doesn't matter who it's by, Jarvis, any friend of yours is a friend of mine. Heave it over and I'll have a look when I return. Now, the binoculars, the stick and the gents' bespoke headwear if you please. I have some intensive marvelling to do.'

Gerin Oil*

GERIN OIL (or Geriniol to give it its scientific name) is a powerful drug which acts directly on the central nervous system to produce a range of symptoms, often of an anti-social or self-damaging nature. It can permanently modify the child brain to produce adult disorders, including dangerous delusions which are hard to treat. The four doomed flights of 11 September 2001 were Gerin Oil trips: all nineteen of the hijackers were high on the drug at the time. Historically, Geriniolism was responsible for atrocities such as the Salem witch-hunts and the massacres of native South Americans by conquistadors. Gerin Oil fuelled most of the wars of the European middle ages and, in more recent times, the carnage that attended the partitioning of the Indian subcontinent and of Ireland.

Gerin Oil intoxication can drive previously sane individuals to run away from a normally fulfilled human life and retreat to closed communities of confirmed addicts. These communities are usually limited to one sex only, and they vigorously, often obsessively, forbid sexual activity. Indeed, a tendency towards agonized sexual

* First published in *Free Inquiry*, December 2003, and then abridged, as 'Opiate of the masses', in *Prospect*, October 2005. I believe it was also translated into Swedish but I can't find the reference. Not sure how they coped with translating 'Gerin Oil' in such a way as to preserve the essential feature of the name. They probably solved the difficulty by leaving it in English.

prohibition emerges as a drably recurring theme amid all the colourful variations of Gerin Oil symptomatology. Gerin Oil does not seem to reduce the libido *per se*, but it frequently leads to a preoccupation with reducing the sexual pleasure of others. A current example is the prurience with which many habitual 'Oilers' condemn homosexuality.

As with other drugs, refined Gerin Oil in low doses is largely harmless, and can serve as a lubricant on social occasions such as marriages, funerals and state ceremonies. Experts differ over whether such social tripping, though harmless in itself, is a risk factor for upgrading to harder and more addictive forms of the drug.

Medium doses of Gerin Oil, though not in themselves dangerous, can distort perceptions of reality. Beliefs that have no basis in fact are immunized, by the drug's direct effects on the nervous system, against evidence from the real world. Oil-heads can be heard talking to thin air or muttering to themselves, apparently in the belief that private wishes so expressed will come true, even at the cost of other people's welfare and mild violation of the laws of physics. This autolocutory disorder is often accompanied by weird tics and hand gestures, manic stereotypies such as rhythmic head-nodding towards a wall, or Obsessive Compulsive Orientation Syndrome (OCOS: facing towards the east five times a day).

Gerin Oil in strong doses is hallucinogenic. Hardcore mainliners may hear voices in the head, or experience visual illusions which seem to the sufferers so real that they often succeed in persuading others of their reality. An individual who convincingly reports high-grade hallucinations may be venerated, and even followed as some kind of leader, by others who regard themselves as less fortunate. Such follower-pathology can long post-date the original leader's death, and may expand into bizarre psychedelia such as the cannibalistic fantasy of 'drinking the blood and eating the flesh' of the leader.

Chronic abuse of Geriniol can lead to 'bad trips', in which the user suffers terrifying delusions, including fears of being tortured, not in the real world but in a postmortem fantasy world. Bad trips of this kind are bound up with a morbid punishment-lore which is as characteristic of this drug as the obsessive fear of sexuality already noted. The punishment culture fostered by Gerin Oil ranges from 'smack' through 'lash' to getting 'stoned' (especially adulteresses and rape victims) and 'demanifestation' (amputation of one hand), up to the sinister fantasy of allo-punishment or 'cross-topping', the execution of one individual for the sins of others.

You might think that such a potentially dangerous and addictive drug would head the list of proscribed intoxicants, with exemplary sentences handed out for pushing it. But no, it is readily obtainable anywhere in the world and you don't even need a prescription. Professional traffickers are numerous and organized in hierarchical cartels, openly trading on street corners and in purpose-made buildings. Some of these cartels are adept at fleecing poor people desperate to feed their habit. 'Godfathers' occupy influential positions in high places, and they have the ear of royalty, of presidents and prime ministers. Governments don't just turn a blind eye to the trade, they grant it tax-exempt status. Worse, they subsidize schools founded with the specific intention of getting children hooked.

I was prompted to write this article by the smiling face of a happy man in Bali. He was ecstatically greeting his death sentence for the brutal murder of large numbers of innocent holidaymakers whom he had never met, and against whom he bore no personal grudge. Some people in the court were shocked at his lack of remorse. Far from remorse, his response was one of obvious exhilaration. He punched the air, delirious with joy that he was to be 'martyred', to use the jargon of his group of abusers. Make no mistake about it: that beatific smile, looking forward with unalloyed pleasure to the firing squad, is the smile of a junkie. Here we have the archetypal mainliner, doped up with hard, unrefined, unadulterated, high-octane Gerin Oil.

Sage elder statesman of the dinosaur fancy*

G REAT HUMORISTS DON'T tell jokes. They plant new species of jokes and then help them evolve, or just sit back and watch them self-propagate, grow, and sprout again. Stephen Potter's *Gamesmanship* is a single elaborated joke, nurtured and sustained through *Lifemanship* and *One-Upmanship*. The joke mutated and evolved with such fertility that, far from fading with repetition, it grew and became funnier. He helped it along by planting supporting memes: 'ploy' and 'gambit', the pseudo-academic footnotes, the fictitious collaborators, Odoreida and Gatling-Fenn – who just *might* not be fictitious. Now, thirty years after Potter's death, if I were to coin, say, Postmodernship, or GM-manship, you would be primed for the joke and ready to go one better. Most Jeeves stories are mutants of one archetypal joke, and again it is a species which evolves and matures to become more funny, not less, with the retelling. The same could be said of *1066 and All That*, *The Memoirs of an Irish RM* and certainly *Lady Addle Remembers*. *How to Keep Dinosaurs* belongs in that great tradition.

Ever since our student days together, Robert Mash has been not just a humorist but a fecund propagator of new evolutionary

---

* Robert Mash is a friend from graduate student days at Oxford. We were fellow members of the Maestro's Mob, the Tinbergen research group. Years later he wrote a lovely book on *How to Keep Dinosaurs*. When it was republished (2003) in a second edition at my instigation, I wrote this foreword.

lineages of humour. If he had a predecessor, it was Psmith: 'That low moaning sound you hear is the wolf bivouacked outside my door' is what I would think of as a Mashian way of saying 'I'm skint.' Psmithian, too, was Mash's grave response to a woman who had just met him at a party. On learning that he was a schoolmaster at a famous school, her innocent conversational question was, 'And do you have girls?' His one-word reply, 'Occasionally,' was calculated to disconcert with exactly Psmith's unblinking solemnity.

Mash's imaginative variants of 'Stap m'vitals' had his whole circle of friends busy inventing new ones, which became ever more bizarre as the species evolved through the memetic microculture. The same for names of pubs. The Rose and Crown in Oxford was our local (where, indeed, much of this early evolution took place) but it was seldom referred to so straightforwardly. 'See you in The Cathedral and Gallbladder' would have been heard somewhere along the evolutionary line. Later specimens seem funny only within the context of their evolutionary history. Another species that Mash planted was the indefinitely evolving variant of the 'Our . . . friend' convolution. To begin with, 'Rose and Crown' might be 'Our floral regal friend' but later descendants of the line evolved the baroque crypticity of a crossword clue and needed a classical education to decipher. The phylum to which all these Mashian species of humour ultimately belonged could be called deadpan circumlocution.

But the youthful Robert Mash as humorist belies the serious scholar of his maturity. Nowhere is his serious side more evident than in this book where he brings together his lifelong expertise on dinosaurs, their habits and maintenance, in sickness and in health. His name has long been a byword in the dinosaur fancy. From show-ring to auction-hall, from racecourse to pterosaur-moor, no gathering of saurophiles is complete until the whisper goes the rounds: 'Mash has arrived.' Even the carnosaurs seem to sense the presence of the master and walk with an added spring to their bipedal step, an added sneer to their bacteria-laced jaws. He is ever ready with a reassuring pat to the diffident hindquarters of a *Compsognathus* or timely advice to its owner.

Is your lap-dinosaur reaching that difficult (not to say uncomfortable) age of needing a spur trim? Mash will advise you on proper pruning before it all ends in tears and inadvertent (and oh so well-meant) laparotomy. Is your gun-dinosaur becoming over-enthusiastic? Call Mash in before it 'retrieves' too many beaters (your retriever's mouth may be as soft as your ghilly's muffled cries for help, but there are limits to both). For those embarrassing moments, as when a *Microraptor* forgets it is in the drawing room, Mash's advice is as discreet as it is succinct. Or are you looking for a load of well-rotted *Iguanodon* manure for the smallholding? Mash is your man.

Though nowadays better known as sage elder statesman of the dinosaur fancy, Robert Mash has seen his share of action. Few who saw him 'up' will forget his insouciant seat on 'Killer', as he nursed that peerless hunter over the twenty-foot jumps to yet another clear round. As for dressage, under 'RM's' spirited martingale even a buck *Brachiosaurus* would prance like a thoroughbred *Ornithomimus*. His view halloo when whipping in that famous pack of twenty *Velociraptor* couple would quicken the pulse of any sportsman, and chill the already cold blood of the hapless *Bambiraptor* gone to ground. And when he donned his well-pounced leather, he was not to be snited at – indeed he was lucratively sought after as a consultant in Arabian royal houses. His freshly enseamed *Pterodactylus*, expertly cast off and with the wind in its sails, would ring up peerlessly before footing and trussing its *Archeopteryx*, with a final, satisfying feake on the gauntlet.

For years, his many friends and admirers on the dinosaur circuit had urged Mash to set down his lifetime's experience in book form, as only he could. The first edition of *How to Keep Dinosaurs* was the result, and it predictably sold out quicker than the whipcrack of an *Apatosaurus* tail. Through the out-of-print wilderness years, well-thumbed bootleg copies became ever more prized possessions, jealously guarded in game bag or Range Rover glove pocket. The need for a second edition became pressing and I am delighted to have been instrumental, however indirectly, in helping to bring it about ('Whoso findeth a publisher findeth a good thing' – Proverbs 18: 22). The second edition has, of course, benefited from Mash's tireless correspondence with dinosaur-owners the world over.

The book can be appreciated on many levels. It is by no means only an owner's manual, though it is indispensably that. For all its sound practical advice, it could only have been written by a professional zoologist, drawing deeply on theory and scholarship. Many of the facts herein are accurate. The world of dinosaurs has always been richly provided with wonder and amazement, and Mash's manual only adds to the mixture. As a theological aside, creationists (now excitingly rebranded as Intelligent Design Theorists) will find it an invaluable resource in their battle against the preposterous *canard* that humans and dinosaurs are separated by sixty-five million years of geological time.

As Robert Mash himself might warn, a dinosaur is for life (very long life in the case of some sauropods), not just for Christmas. The same could be said of his book. Nevertheless, it will make a delightful present for anyone, of any age, and for many Christmases to come.

Athorism: let's hope it's a lasting vogue*

A THORISM IS ENJOYING a certain vogue right now. Can there be a productive conversation between Valhallans and athorists? Naive literalists apart, sophisticated thoreologians long ago ceased believing in the material substance of Thor's mighty hammer. But the spiritual essence of hammeriness remains a thunderingly enlightened revelation, and hammerological faith retains its special place in the eschatology of neo-Valhallism, while enjoying a productive conversation with the scientific theory of thunder in its non-overlapping magisterium. Militant athorists are their own worst enemy. Ignorant of the finer points of thoreology, they really should desist from their strident and intolerant strawmandering, and treat Thor-faith with the uniquely protected respect it has always received in the past. In any case, they are doomed to failure. People need Thor, and nothing will ever remove him from the culture. What are you going to put in his place?

---

* The *Washington Post* used to have a regular feature called 'On Faith', moderated by Sally Quinn, to which I was a frequent contributor. This is the opening paragraph of a piece that appeared on New Year's Day, 1 January 2007, in response to a question on the current vogue for atheism.

AFTERWORD

This joke could run and run. Feminist thoreologians prefer to downplay the patriarchally hard phallic aspects of Thor's hammer, liberation thoreologians find common cause with workers marching under the banner of the hammer and sickle, while for postmodern thoreologists, the hammer is a puissant significator of deconstruction. Continue to taste.

Dawkins' Laws*

Dawkins' Law of the Conservation of Difficulty

Obscurantism in an academic subject expands to fill the vacuum of its intrinsic simplicity.

Dawkins' Law of Divine Invulnerability

God cannot lose.

Lemma 1: When comprehension expands, gods contract – but then redefine themselves to restore the status quo.

Lemma 2: When things go right, God will be thanked. When things go wrong, he will be thanked that they are not worse.

Lemma 3: Belief in the afterlife can only be proved right, never wrong.

Lemma 4: The fury with which untenable beliefs are defended is inversely proportional to their defensibility.

---

* This was my response to the question 'What is your Law?', posed by John Brockman in 2004 as his annual challenge sent around to the members of his online salon *The Edge*: https://www.edge.org/annual-question/whats-your-law.

Dawkins' Law of Hell and Damnation

$$H \propto 1/P$$

where H is the threatened temperature of hell fire and P is the perceived likelihood that it exists.

Or, in words: 'The magnitude of a threatened punishment is inversely proportional to its plausibility.'

The following law, though probably older, is often attributed to me in various versions, and I am happy to formulate it here as:

The Law of Adversarial Debate

When two incompatible beliefs are advocated with equal intensity, the truth does not necessarily lie halfway between them. One side can be simply wrong.

VIII

NO MAN IS AN ISLAND

FROM NEWTON'S 'standing on the shoulders of giants' and before, science has always been a collaborative venture. While it would be very unDawkinsian panglossianism to deny that some of its practitioners have insufficiently acknowledged what their work owes to the contributions of others, many, many more epitomize that collegiality, cooperativeness of spirit and mutual respect which the first piece in this collection identified as among the chief 'values of science'. These values, of course, enriched by personal attachment and moral sensibility, are those not of scientists alone but of civilized humanity. They are celebrated in this final brief section, which presents a small selection of personal reflections in memory and honour of others.

'Memories of a maestro' was originally delivered as the opening address to a conference gathered in memory of the Nobel Prize-winning biologist Niko Tinbergen. It speaks not only of professional regard but of the sense of belonging generated by participation in the shared endeavour of learning and exploration, the privilege attached to membership not simply of an elite institution but of a group of individuals as gifted in teaching as in their pursuit of science. It speaks, too, of the deeply felt obligation to continue this cascading of knowledge through future generations: 'We . . . wanted people to pick up the torches that Niko had passed them, and run on with them towards the future.'

The next two pieces, 'O my beloved father' and 'More than my uncle', shine with pride in, and love for, family past and present. Where a less scrupulously honest son and nephew of left-liberal inclinations might have been tempted to downplay, gloss over or repudiate a cast-iron imperial heritage, Richard will have no truck with such weaselling, in either direction: 'There was, of course, quite a bit that was bad about the British in Africa. But the good was very very good, and Bill was one of the best.' These affectionate remembrances are typically lit by humour, as in his accounts of his Uncle Bill's determined reading of the Riot Act ('I imagine the text as sewn into the lining of his pith helmet') and his father's 'Heath Robinson' inventiveness on the family farm. And they resonate with pride as much in unashamed paternal and avuncular love as in any of his forebears' (considerable) worldly achievements: 'Air of command and military bearing be blowed. There are greater qualities to admire.'

Readers of this collection will, I hope, have come to appreciate the huge scope of Richard Dawkins' preoccupations, passions and talents – as a scientist, a teacher, a polemicist, a humorist, above all as a writer. For the final piece in this volume, 'Honouring Hitch', I have chosen one that focuses this dazzling versatility to a single brilliant point. This address, given by Richard in presenting the award made in his name by the Atheist Alliance of America to the then mortally ill Christopher Hitchens, resounds with, as he says, 'admiration, respect, and love'. It's a curious but fitting irony that many of the tributes he pays to Hitchens could equally justly be paid to himself: 'the leading intellect and scholar of our atheist/ secular movement'; a 'gently encouraging friend to the young, to the diffident'; capable equally of 'the penetratingly logical', 'the cuttingly witty' and 'the courageously unconventional'. No wonder they were soul brothers.

Richard Dawkins will always have his critics – some sympathetic to his aims, some deeply hostile. But an honest reader of any stamp will, I think, find it hard to deny that 'there was quite a bit that was bad about British writing in our times. But the good was very very good, and Richard Dawkins was one of the best.'

G.S.

Memories of a maestro*

WELCOME TO OXFORD. For many of you it is welcome back to Oxford. Perhaps even, for some of you, it would be nice to think that it might feel like welcome home to Oxford. And it is a great pleasure to welcome so many friends from the Netherlands.

---

* Niko Tinbergen, who shared the 1973 Nobel Prize in Physiology with Konrad Lorenz and Karl von Frisch, had been lured to Oxford from his native Holland in 1949. He accepted the invitation partly (but only partly, according to Hans Kruuk's highly perceptive and honest biography) because he saw Oxford as a springboard from which to take Dutch and German ethology to the English-speaking world. The move involved considerable personal sacrifice. He voluntarily took a substantial cut in salary and a demotion from full professor at Leiden to 'demonstrator', the lowest rank in Oxford's academic hierarchy; his children had to take a crash course in English to cope with (expensive) new schools; and he never found the Oxford college system congenial. British academic biology was lucky to get him. I arrived in his research group in 1962, perhaps a little too late to benefit fully from his heyday, but I got plenty of it at second hand from the large and flourishing group that he founded and influenced, above all Mike Cullen, to whom I paid tribute in *An Appetite for Wonder*. A year after Niko died, Marian Stamp Dawkins, Tim Halliday and I organized a memorial conference in Oxford. What follows is my opening speech, which served as the intro-duction to the proceedings of the conference which we edited as a book, *The Tinbergen Legacy*.

Last week, when everything had been settled except final, last-minute arrangements, we heard that Lies Tinbergen had died. Obviously we would not have chosen such a time to have this meeting. I'm sure we'd all like to extend our deep sympathy to the family, many of whom, I'm happy to say, are present. We discussed what we should do and decided that, in the circumstances, there was nothing for it but to carry on. The members of the Tinbergen family that we were able to consult were fully in agreement. I think we all knew that Lies was an enormous support to Niko, but I think that very few of us really knew how much of a support she was to him, particularly during the dark times of depression.

I should say something about this memorial conference and what led up to it. People have their own ways of grieving. Lies's way was to take literally Niko's characteristically modest instruction that he wanted no funeral or memorial rites of any kind. There were those of us who were fully sympathetic to the desire for no religious observance, but who nevertheless felt the need for some kind of rite of passage for a man whom we had loved and respected for so many years. We suggested various kinds of secular observance. For instance, the fact that there was such musical talent in the Tinbergen family led some of us to suggest a memorial chamber concert with readings or eulogies in the intervals. Lies made it very clear, however, that she wanted nothing of the kind and that Niko would have felt the same.

So we did nothing for a while. Then, after some time had elapsed, we realized that a memorial conference would be sufficiently different from a funeral as not to count. Lies accepted this, and there came a time, during our planning of the conference, when she said that she hoped to attend the conference, although she later changed her mind about that, thinking, again with characteristic modesty and completely erroneously, that she would have been in the way.

It is an enormous pleasure to welcome so many old friends. It is a tribute to Niko, and the affection that his old pupils felt for him, that so many of you are here today, converging on Oxford from, in some cases, very far away. The list of people coming is a galaxy of old friends, some of whom may not have set eyes on one another for thirty years. Just reading the guest list was a moving experience for me.

We shall all of us have memories of Niko and of the group of his associates with whom we happen to be contemporary. My own begin when I was an undergraduate and he lectured to us, not at first on animal behaviour but on molluscs – for it was Alister Hardy's quaint idea that all the lecturers should participate in the 'Animal Kingdom' course which is one of the sacred cows of Oxford zoology. I didn't know, then, what a distinguished man Niko was. I think that if I had, I'd have been rather aghast at his being made to lecture on molluscs. It was bad enough that he gave up being a professor in Leiden to become, by Oxford's snobbish custom, just plain 'Mr Tinbergen'. I don't remember much from those early mollusc lectures, but I do remember responding to his wonderful smile: friendly, kindly, avuncular as I thought then, although he must have been scarcely older than I am now.

I think I must have been imprinted on Niko and his intellectual system then, for I asked my college tutor if I could have tutorials with Niko. I don't know how he managed to swing it, because I don't think Niko gave undergraduate tutorials as a rule. I suspect that I may have been the last undergraduate to have had tutorials with him. Those tutorials had an enormous influence on me. Niko's style as a tutor was unique. Instead of giving a reading list with some sort of comprehensive coverage of a topic, he would give a single, highly detailed piece of work, such as a DPhil thesis. My first one, I remember, was a monograph by A. C. Perdeck, who I am happy to say is here today. I was asked simply to write an essay on anything that occurred to me as a result of reading the thesis or monograph. In a sense it was Niko's way of making the pupil feel like an equal – a colleague whose views on research were worth hearing, not just a student mugging up a topic. Nothing like this had ever happened to me before, and I revelled in it. I wrote huge essays that took so long to read out that, what with Niko's frequent interruptions, they were seldom finished by the end of the hour. He strode up and down the room while I read my essay, only occasionally coming to rest on whatever old packing case was serving him as a chair at the time, chain-rolling cigarettes and obviously giving me his whole attention in a way that, I'm sorry to say, I cannot claim to do for most of my pupils today.

As a result of these marvellous tutorials, I decided that I very much wanted to do a DPhil with Niko. And so I joined the 'Maestro's Mob', and it was an experience never to be forgotten. I remember with particular affection the Friday evening seminars. Apart from Niko himself, the dominant figure at that time was Mike Cullen. Niko obstinately refused to let sloppy language pass, and proceedings could be stalled for an indefinite period if the speaker was not able to define his terms with sufficient rigour. These were arguments in which everybody became engaged, eager to make a contribution. If, as a result, a seminar wasn't finished at the end of the two hours, it simply resumed the following week, no matter what might have been previously planned.

I suppose it may have been just the naivety of youth, but I used to look forward to those seminars with a sort of warm glow for the whole week. We felt ourselves members of a privileged elite, an Athens of ethology. Others, who belonged to different cohorts, different vintages, have talked in such similar terms that I believe that this feeling was a general aspect of what Niko did for his young associates.

In a way, what Niko stood for on those Friday evenings was a kind of ultra-rigorous, logical commonsense. Put like that, it may not sound like much; it may seem even obvious. But I have since learned that rigorous commonsense is by no means obvious to much of the world. Indeed, commonsense sometimes requires ceaseless vigilance in its defence.

In the world of ethology at large, Niko stood for breadth of vision. He not only formulated the 'four questions' view of biology, he also assiduously championed any one of the four that he felt was being neglected. Since he is now associated in people's minds with field studies of the functional significance of behaviour, it is worth recalling how much of his career was given over to, for instance, the study of motivation. And, for what it is worth, my own dominant recollection of his undergraduate lectures on animal behaviour was of his ruthlessly mechanistic attitude to animal behaviour and the machinery that underlay it. I was particularly taken with two phrases of his – 'behaviour machinery' and 'equipment for survival'. When I came to write my own first book, I combined them into the brief phrase 'survival machine'.

In planning this conference, we obviously decided to concentrate on fields that Niko had been pre-eminent in, but we didn't want the talks to be only retrospective. Of course we wanted to spend some time looking back at Niko's achievements, but we also wanted people to pick up the torches that Niko had passed them, and run on with them towards the future.

Torch-running behaviour, in new and exciting directions, bulks so large in the ethograms of Niko's students and associates that planning the programme was a major headache. 'How on earth', we asked ourselves, 'can we possibly leave out so-and-so? On the other hand, we have space for only six talks.' We could have limited ourselves to Niko's own pupils – his scientific children, but this would have been to devalue his enormous influence via grandpupils and others. We could have concentrated on people and major areas not covered in the *Festschrift* volume edited by Gerard Baerends, Colin Beer and Aubrey Manning, but that too would have been a pity. In the end, it seemed almost not to matter which half a dozen of Niko's intellectual descendants stood up to represent the rest of us. And perhaps that is the true measure of his greatness.

O my beloved father:
John Dawkins, 1915-2010*

MY FATHER, CLINTON JOHN DAWKINS, who has died peacefully of old age, lived his ninety-five years to the full and packed an enormous amount into them.

He was born in Mandalay in 1915, the eldest of three talented brothers, all of whom were to follow their father and grandfather into the colonial service. John's boyhood hobby of pressing flowers, reinforced by a famous biology teacher (A. G. Lowndes of Marlborough), led him to read botany at Oxford, and thence to study tropical agriculture at Cambridge and ICTU (Trinidad) in preparation for posting to Nyasaland as a junior agricultural officer. Immediately before leaving for Africa, he married my mother, Jean Ladner. She followed him soon afterwards and they began an idyllic married life at various remote agricultural stations before he was called up for wartime service in the King's African Rifles (KAR). John wangled permission to travel up to Kenya under his own steam rather than with the regimental convoy, which enabled Jean

---

* I hope it will not be seen as self-indulgence to include two family memorials. They are not directly connected with science but, in the sense in which I can be said to have a soul, they are connected with mine. My father and his two brothers all influenced me in their different ways. This first piece is the obituary which I published in the *Independent*, 11 December 2010.

to accompany him – illegally, which I guess illegitimizes my own birth in Nairobi.*

John's post-war work as an agricultural officer back in Nyasaland was interrupted when he received an unexpected legacy from a very distant cousin. Over Norton Park had been owned by the Dawkins family since the 1720s, and Hereward Dawkins, casting around the family tree for a Dawkins heir, could find none closer than the young agricultural officer in Nyasaland, whom he had never met and who had never heard of him.

Hereward's gamble paid off in spades. The young couple decided to leave Africa and run Over Norton Park as a commercial farm rather than as a gentleman's estate. Against great odds (and discouraging advice from family and family solicitor) they succeeded, and they could fairly be said to have saved the family inheritance.

They turned the big house into flats, specializing in colonial servants sent 'home' on leave. Tractors didn't have cabs in those days, and John, wearing his old KAR hat (think Australian bushwhacker) could be heard across two fields bellowing the psalms at the top of his voice ('Moab was my washpot') on his diminutive Ferguson tractor (just as well it was diminutive, since he once contrived to run himself over with it).

Equally diminutive were the Jersey cows that graced the parkland. Their (now unfashionably) rich milk was separated into cream, which supplied most of the Oxford colleges and lots of shops and restaurants, while, in a neat display of what John called 'music and movement', the skim milk nourished the large herd of Over Norton pigs. The cream separation itself involved a virtuoso display of John's characteristic Heath Robinson† ingenuity, lashed up with

---

* Her journal of this journey and her subsequent life as an army camp follower in Kenya and Uganda makes entertaining reading, and I quoted passages from it in my first memoir, *An Appetite for Wonder*.

† For the American equivalent, think Rube Goldberg.

binder twine – the inspiration for a wonderful verse composed by the long-serving pig-man: 'With clouds of steam and lights that flash, / the scheme is most giganto, / When churns take wings on nylon slings / Like fairies at the panto.'

John didn't limit his binder-twine ingenuity to his farming activities. Throughout his life he took up one creative hobby after another, and they all benefited from his resourcefulness with red string and dirty old scrap metal. Each Christmas there would be a new crop of home-made presents, beginning with the toys he made for me and my sister in Africa, and moving on to equally beguiling presents for grandchildren and great-grandchildren.

He was elected a Fellow of the Royal Photographic Society, his special art form being the use of two projectors to 'dissolve' pictures into carefully matched images in sequence. Each sequence had a theme, and his themes ranged from autumn leaves, through his beloved Ireland, to abstract art created by photographing the spectral patterns lurking in the deep interiors of cut-glass decanter stoppers. He automated the dissolving process by making his own 'iris diaphragms' for the alternating projectors, held together with rubber bands. Inexpensive and very effective.[*]

In his nineties John slowed down and his memory slipped away. But he accepted old age with the same generous grace as had attended his active years. He and Jean, who survives him, celebrated their seventieth wedding anniversary last year in a splendid family party. He learned to laugh at his infirmities with a benign cheerfulness that inspired deep love in their large extended family, including nine great-grandchildren, living in four separate houses all within the dry stone Cotswold walls of Over Norton Park – the ancestral home that he and Jean[†] had saved.

---

[*] Nowadays, of course, it would be done by computer.

[†] She celebrated her hundredth birthday a few days before I write this footnote.

More than my uncle: A. F. 'Bill' Dawkins, 1916–2009*

IN 1972, THE British government was trying to find a solution to the problem that was then Rhodesia. The Foreign Secretary, Sir Alec Douglas-Home, appointed a Royal Commission, under Lord Pearce, to tour the villages and byways of Rhodesia, trying to canvass popular opinion. The commissioners were old colonial types who, it was rightly supposed, had the necessary experience. Bill Dawkins was a natural for the Pearce Commission, and he was duly called out of retirement.

At that time, my Oxford college had an ancient and garrulous old classics don, living in, who had spent much of his life closely associated with the colonial service. Sir Christopher became obsessed with the Pearce Commission, and especially obsessed with Bill, probably because the BBC had taken to using his handsome features as their icon for that item on the news each night. As Lalla might have put it, Bill was excellent casting for the role. Although he had never met Bill, Sir Christopher clearly felt he knew him, as a kind of epitome of imperial uprightness and strength of character. This showed itself in such remarks as, 'Dawkins' uncle

---

* My father's middle brother Bill predeceased him by a year. I delivered this eulogy for my dear uncle (and godfather) at his funeral at St Michael and All Angels Church, Stockland, Devon, on Wednesday, 11 November 2009. Since it was a family funeral I obviously referred to members of his family by their first names without explanation.

would soon put a stop to *that*.' Or: 'I'd like to see anyone try to pull a fast one on Dawkins' uncle. *Ha!*'

The Pearce Commissioners were sent out to tour the country in pairs, with an entourage, and Bill was paired with another old colonial called Burkinshaw. True to Bill's iconic status, the BBC news cameras chose to follow Dawkins and Burkinshaw on one of these fact-finding missions, and Sir Christopher was agog in front of the television screen. I vividly remember his summation, the next day, in his distinctive old raconteur's voice: 'About Burkinshaw I will say nothing. *Dawkins*, however, is obviously accustomed to *commanding men*.'

David Attenborough told me he had exactly the same impression of Bill, and he drew himself up to his full height and pulled a realistically imperious face to illustrate the point. He had stayed with Bill and Diana while on a filming trip to Sierra Leone in 1954, and they remained friends thereafter.

I can't imagine anybody ever calling Bill either Arthur or Francis, although A.F. suited him well enough. Throughout his life, he was never called anything but Bill, which dated from babyhood when he was said to resemble Bill the Lizard in *Alice in Wonderland*. I looked up to him from the first day I met him. It was 1946, I was five years old and in the bath in the family house at Mullion. Bill must have just arrived from Africa, and my father brought his younger brother in to see me. I was awed by this tall, handsome figure, with black hair and moustache, blue eyes, and a strong military bearing. I looked up to him throughout my life, as a shining example of all that was good about the British in Africa. There was, of course, quite a bit that was bad about the British in Africa. But the good was very very good, and Bill was one of the best.

He was a notable athlete. At the prep school that I attended some twenty-five years after him, I remember my family pride at seeing his name on the roll of honour as the holder of the school record for the one hundred yards. This speed obviously stood him in good stead later when, in the early stages of the war, he played rugby for the Army. I managed to track down a report of 22 April 1940, from the *Times* rugby correspondent, of what must have been an exciting match between the army and the Great Britain team, which the army won. Late in the game, it transpired that:

> The Army passing remained ragged, but Dawkins and Wooller, by sheer dash and ability to pick up on the run, soon reminded Great Britain that these two players alone would take a lot of stopping given half a chance. First, Dawkins, at a great pace, sent Wooller striding for the line, with a stupendous dive-over at the finish. Next, Wooller sent in Dawkins.

Evidently the speed that won Bill the school record for the hundred-yard dash had not deserted him, and 'dash' was clearly still the right word. 'Great pace', 'sheer dash' and 'obviously accustomed to commanding men . . .' But these phrases, impressive as they are, may represent the least of the qualities that we remember today. Here is a letter from a gentle and loving father, to six-year-old Penny.

> Do you remember the Morning Glory outside the house and sometimes we used to count the flowers on my way to the office and the most we ever got to was 54. Well, today there were 91 all on one side. Have you read this all without any help, because I have not used any long words like ANTIDISESTABLISHMENTARIANISM, HAVE I? . . . Lots of Love XXXX from Daddy

I know people who would have given their eye teeth for a father like that, let alone a stepfather.

Bill was born in Burma in 1916. While his parents were still out there, he and his elder brother John were sent to boarding school in England, and spent their holidays with grandparents here in Devon, which is presumably when he acquired his love of this beautiful county.

By coincidence he was later to find himself back in Burma for the whole of the war, fighting the Japanese as an officer in the Sierra Leone Regiment, for it was British practice to use tropical soldiers in tropical theatres of war. He rose to the rank of major, and was mentioned in despatches.*

He came to love the Sierra Leone people through commanding them in war; and after the war, when he followed the khaki-shorted family tradition of joining the colonial service, he applied to go to Sierra Leone, where he was promoted to district commissioner in 1950.

It was a tough job, and he occasionally had to quell disturbances and riots, armed with nothing more than his innate air of being 'accustomed to commanding men'. The riots were not aimed at the colonial government but were to do with fighting between rival tribes. Bill, the district commissioner, went striding in and read the Riot Act. Not *metaphorically* read the Riot Act: he *literally* read the Riot Act, every word of it. (I imagine the text as sewn into the lining of his pith helmet.) During one riot, Bill picked up an injured man and carried him to safety. The rioters tried to persuade him to put the man down, so that they could continue beating him up. Bill refused, knowing that, so long as he was carrying him, they wouldn't dare to hurt him. This curiously surreal approach to rioting reached its climax when, in the middle of one riot, everything suddenly went quiet as somebody shouted that 'The DC he done tire', and a table and chair were lowered on a rope, from an upper window. According to Penny, who told me this story, a bottle of beer was solemnly placed on the table, and Bill was invited to sit down and drink the beer. This he did. Whereupon the table and chair were hauled upstairs again, and the riot resumed as if nothing had happened.

During another riot, one of the Africans was heard to shout these words of reassurance to everyone who could hear above the hubbub: 'It's all right everybody, everything will soon be all right, Major Donkins has arrived.' Presumably this was said by one of his wartime soldiers from Burma days, because Bill would never have

---

* My mother, who was close to her brother-in-law (two ways, for the two brothers married two sisters), recently told me Bill would never talk about his wartime experiences. No wonder, given where and how he had spent those years.

used his military rank in peacetime. His name *was* widely mispronounced in Sierra Leone as Donkins. And on a later occasion, a letter addressed to 'The Colonial Donkey, Freetown', was successfully delivered.

Here's another letter to Bill from that period, dated 22 November 1954. It has nothing to do with riots, but is a letter of farewell from a grateful African (with an agenda). It read as follows:

22nd November 1954

My Dear Sir,

Farewell faithful friend, I must now bid adieu to these joys and pleasures I have tasted with you. We have laboured together united in heart but now we must close and soon we must part. My heart sinked within me to bid you adieu. Though absent in body I am with you in prayer that I will meet and work under you some where some how.

As the dearest friend of Mankind that is Jesus gave his body and blood as token and rememberance to his disciples that they remember him, so also I want you to give a token and that is a Permitt to purchase a single barrel shot gun . . .

It is always hard to make a new acquaintance. If I therefore leave the matter untouched it will then take some years. This matter however is suitable for this occasion as it will be a rememberance. I shall remember you through the Gun.

With every respect and honour to you Sir

I am

Your Obedient Servant

Self-serving though this letter may be, the affection and respect shine through, and we may be sure that that part, at least, was sincere.

Bill's success as a DC was recognized in 1956, when he received an unexpected and rather glamorous promotion: seconded to run the West Indian island of Montserrat. The whole family moved to Government House on this tiny island, where Bill was, not quite literally, monarch of all he surveyed. It was then a paradise, before the catastrophes of Hurricane Hugo and the terrible volcano eruption that laid waste to the island, where Thomas and Judith still loyally soldier on. Bill was the Queen's Official Representative, so they had the Crown on their car instead of an ordinary number plate, and a flag on the bonnet, which was unfurled only when 'His Honour' was actually in the car. Diana played the role of consort, and we may be sure she played it to the full: Patron of the Girl Guides, opening fetes and bazaars, and lots more. It must have seemed very different from the jungles of Sierra Leone. And Diana would have been brilliant at it, as she was in all other aspects of their life together. Bill played cricket for Montserrat against other West Indian islands, and was actually quite badly injured while keeping wicket.

Following the Montserrat interlude, when Bill's secondment came to an end, he was offered another West Indian Island, Grenada, but instead he characteristically opted to return to Africa, where the challenge was tougher and the need greater. He went back to Sierra Leone, now raised to the rank of provincial commissioner. At the end of this period, when Sierra Leone gained independence, he was again offered a West Indian island: Governor of St Vincent. As a full governorship, this would have carried a knighthood. However, mindful that his father, my grandfather, was ageing, and that Penny, at Cambridge, and Thomas, at Marlborough, might need a home base in England, he and Diana decided that he would retire from the colonial service and take a job as a schoolmaster.

He had read mathematical mods at Balliol, and so was equipped to teach mathematics. This he did, with great success, at Brentwood School. His dark good looks must, by then, have matured into something more formidable, for his nickname at Brentwood was Dracula. Or perhaps this was just a reference to his ability to keep order in class, a quality that is not universal among schoolmasters. Yet again, he was 'accustomed to commanding men'.

Air of command and military bearing be blowed. There are greater qualities to admire. Bill was a loving husband, brother, father, grandfather and ... uncle. Uncle Bill was more than my uncle, he was my godfather. In later life he laughingly said *failed* godfather, but with hindsight I think he did take a more than merely avuncular interest in my welfare. Either that, or he was just immensely kind to everyone. Which, now that I think about it, he was.

Towards the end of his life, he gave me one godfatherly piece of advice. He probably said it to others, but when he said it to me it was with a piercing look in those blue eyes, filled with wisdom and experience, which told me this was going to be a serious warning for a godson. 'You do know, don't you? Old age is a *bugger*.'

Well, he is liberated from that now, and at peace. He may have been accustomed to commanding men, but he was loved by them too. He was loved by everyone who knew him. He left the world a better place than he found it – several different places around the world. We mourn him. But, at the same time, we rejoice in him, and what he has left behind.

AFTERWORD

My father's youngest brother, Colyear, was academically the cleverest of the three. I didn't have the opportunity to write his obituary, but I dedicated *River Out of Eden* to the memory of 'Henry Colyear Dawkins (1921–1992), Fellow of St John's College, Oxford: a master of the art of making things clear'. Two anecdotes are worth adding here by way of illuminating his character. One is taken from the obituary by his forester colleague Robert Plumptre. While on a troopship in the war, somewhere in the Indian Ocean, Colyear constructed a home-made sextant in order to discover where they were (which the soldiers were not allowed to know, for security reasons). The instrument was confiscated and he was briefly suspected of being a spy.

For the second, which similarly calls to mind the dundridge mentality of officialdom against which I inveighed earlier,* I quote from my memoir, *Brief Candle in the Dark*:

---

* 'If I ruled the world . . .', page 321.

At Oxford railway station the car park was guarded by a mechanical arm, which rose to allow each car to leave when the driver inserted a token of payment in a slot. One night Colyear had returned to Oxford on the last train from London. Something had gone wrong with the mechanism of the arm and it was stuck in the down position. The station officials had all gone home, and the owners of the trapped cars were in despair as to how to escape the car park. Colyear, with his bike waiting, had no personal interest; nevertheless, with exemplary altruism he seized the arm, broke it, carried it up to the stationmaster's office and plonked it down outside the door with a note giving his name, address and explanation as to why he had done it. He should have been given a medal. Instead, he was prosecuted in court and fined. What a terrible incentive to public-spiritedness. How very typical of the rule-obsessed, legalistic, mean-spirited dundridges of today's Britain.

And a little sequel to that story. Many years later, after Colyear's death, I chanced to meet the distinguished Hungarian scientist Nicolas Kurti (a physicist who incidentally happened to be a pioneer of scientific cookery, injecting meat with a hypodermic syringe, all that sort of thing). His eyes lit up when I spoke my name.

'Dawkins? Did you say Dawkins? Are you any relation of the Dawkins who broke the arm at the Oxford station car park?'

'Er, yes, I'm his nephew.'

'Come, let me shake your hand. Your uncle was a hero.'

If the magistrates who imposed Colyear's fine should happen to read this, I hope you feel thoroughly ashamed. You were only doing your duty and upholding the law? Yeah, right.

412

Honouring Hitch*

Today I am called upon to honour a man whose name will be joined, in the history of our movement, with those of Bertrand Russell, Robert Ingersoll, Thomas Paine, David Hume.

* Christopher Hitchens died of cancer in December 2011. Two months earlier I travelled to Houston, Texas and conducted a long interview with him for the *New Statesman*. I think it was the last major interview he gave. I had been invited to edit the Christmas issue of the magazine, and this interview was one of the main features of 'my' issue (another was 'The tyranny of the discontinuous mind': see page 287). The day after the interview, he attended the Texas Freethought Convention in Houston. In 2003 the Atheist Alliance of America had introduced an annual award, the Richard Dawkins Award, to honour those who raise public consciousness of atheism. I play no part in the annual choice of recipient, but I'm usually invited to present it at a conference, in person or by video. And I feel hugely honoured myself by every one of the illustrious names on the list, now fourteen strong. In 2011 the award went to Christopher Hitchens, and it was to be presented to him at the Texas Freethought Convention. He was too weak to attend most of the conference, but he entered towards the end of the banquet, to a thunderous and highly emotional standing ovation. I then made the speech reproduced here. At the end of it he came up on the platform, we embraced, and he made a speech of his own. His voice was weak and interrupted by fits of coughing, but it was a *tour de force* from a valiant fighter, the finest orator I ever heard. He even found enough stamina to take a large number of questions at the end. It is a privilege to have known him. I wish I had known him better.

He is a writer and an orator with a matchless style, commanding a vocabulary and a range of literary and historical allusion wider than anybody I know. And I live in Oxford, his alma mater and mine.

He is a reader, whose breadth of reading is simultaneously so deep and comprehensive as to deserve the slightly stuffy word 'learned' – except that Christopher is the least stuffy learned person you will ever meet.

He is a debater, who will kick the stuffing out of a hapless victim, yet he does it with a grace that disarms his opponent while simultaneously eviscerating him. He is emphatically not of the (all too common) school that thinks the winner of a debate is he who shouts loudest. His opponents may shout and shriek. Indeed they do. But Hitch doesn't need to shout. His words, his polymathic store of facts and allusions, his commanding generalship of the field of discourse, the forked lightning of his wit . . . I tried to sum it up in my review of *God is not Great* in the *Times* of London:

> There is much fluttering in the dovecots of the deluded, and Christopher Hitchens is one of those responsible. Another is the philosopher A. C. Grayling. I recently shared a platform with both. We were to debate against a trio of, as it turned out, rather half-hearted religious apologists ('Of course I don't believe in a God with a long white beard, but . . .'). I hadn't met Hitchens before, but I got an idea of what to expect when Grayling emailed me to discuss tactics. After proposing a couple of lines for himself and me, he concluded, '. . . and Hitch will spray AK47 ammo at the enemy in characteristic style'.

Grayling's engaging caricature misses Hitchens' ability to temper his pugnacity with old-fashioned courtesy. And 'spray' suggests a scattershot fusillade, which underestimates the deadly accuracy of his marksmanship. If you are a religious apologist invited to debate with Christopher Hitchens, decline. His witty repartee, his ready-access store of historical quotations, his bookish eloquence, his effortless flow of well-formed and beautifully spoken words, would threaten your arguments even if you had good ones to deploy. A string of reverends and 'theologians' ruefully discovered this during Hitchens' barnstorming book tour around the United States.

With characteristic effrontery, he took his tour through the Bible Belt states – the reptilian brain of southern and middle America, rather than the easier pickings of the country's cerebral cortex to the north and down the coasts. The plaudits he received were all the more gratifying. Something is stirring in this great country.

Christopher Hitchens is known as a man of the left. Except that he is too complex a thinker to be placed on a single left–right dimension. Parenthetically, I have long been surprised that the very idea of a single left–right political spectrum works at all. Psychologists need many mathematical dimensions in order to locate human personality, and why should political opinion be any different? With most people, it is surprising how much of the variance is explained by the single dimension we call left–right. If you know somebody's opinion on, say, the death penalty, you can usually guess their opinion on taxation or public health.

But Christopher is a one-off. He is unclassifiable. He might be described as a contrarian except that he has specifically and correctly disavowed the title. He is uniquely placed in his own multi-dimensional space. You don't know what he will say about anything until you hear him say it, and when he does he will say it so well, and back it up so fully, that if you want to argue against him you'd better be on your guard.

He is known throughout the world as one of the leading public intellectuals anywhere. He has written many books and countless articles. He is an intrepid traveller and a war reporter of signal valour.

But of course he has a special place in our affections here as the leading intellect and scholar of our atheist/secular movement. A formidable adversary to the pretentious, the woolly-minded or the intellectually dishonest, he is a gently encouraging friend to the young, to the diffident, to those tentatively feeling their way into the life of the freethinker and not certain where it will take them.

We treasure his *bons mots* and I'll just quote a few of my favourites.

From the penetratingly logical . . .

That which can be asserted without evidence, can be dismissed without evidence.

To the cuttingly witty:

Everybody does have a book in them, but in most cases that's where it should stay.

To the courageously unconventional:

[Mother Teresa] was not a friend of the poor. She was a friend of poverty. She said that suffering was a gift from God. She spent her life opposing the only known cure for poverty, which is the empowerment of women and the emancipation of them from a livestock version of compulsory reproduction.

The following is vintage Hitch:

I suppose that one reason I have always detested religion is its sly tendency to insinuate the idea that the universe is designed with 'you' in mind or, even worse, that there is a divine plan into which one fits whether one knows it or not. This kind of modesty is too arrogant for me.

And what about this:

Organised religion is violent, irrational, intolerant, allied to racism, tribalism, and bigotry, invested in ignorance and hostile to free inquiry, contemptuous of women and coercive toward children.

And this:

Everything about Christianity is contained in the pathetic image of 'the flock'.

His respect for women and their rights shines forth:

Who are your favorite heroines in real life? The women of Afghanistan, Iraq, and Iran who risk their lives and their beauty to defy the foulness of theocracy.

Though not a scientist and with no pretensions in that direction, he understands the importance of science in the advancement of our species and the destruction of religion and superstition:

> One must state it plainly. Religion comes from the period of human prehistory where nobody – not even the mighty Democritus who concluded that all matter was made from atoms – had the smallest idea what was going on. It comes from the bawling and fearful infancy of our species, and is a babyish attempt to meet our inescapable demand for knowledge (as well as for comfort, reassurance and other infantile needs). Today the least educated of my children knows much more about the natural order than any of the founders of religion.

He has inspired and energized and encouraged us. He has us cheering him on almost daily. He's even begotten a new word – the hitchslap. We don't just admire his intellect, we admire his pugnacity, his spirit, his refusal to countenance ignoble compromise, his forthrightness, his indomitable spirit, his brutal honesty.

And in the very way he is looking his illness in the eye, he is embodying one part of the case against religion. Leave it to the religious to mewl and whimper at the feet of an imaginary deity in their fear of death; leave it to them to spend their lives in denial of its reality. Hitch is looking it squarely in the eye: not denying it, not giving in to it, but facing up to it squarely and honestly and with a courage that inspires us all.

Before his illness, it was as an erudite author and essayist, a sparkling, devastating speaker that this valiant horseman led the charge against the follies and lies of religion. Since his illness he has added another weapon to his armoury and ours – perhaps the most formidable and powerful weapon of all: his very character has become an outstanding and unmistakable symbol of the honesty and dignity of atheism, as well as of the worth and dignity of the human being when not debased by the infantile babblings of religion.

Every day he is demonstrating the falsehood of that most squalid of Christian lies: that there are no atheists in foxholes. Hitch is in a foxhole, and he is dealing with it with a courage, an honesty and a dignity that any of us would be, and should be, proud to muster. And in the process, he is showing himself to be even more deserving of our admiration, respect, and love.

I was asked to honour Christopher Hitchens today. I need hardly say that he does me the far greater honour, by accepting this award in my name. Ladies and gentlemen, comrades, I give you Christopher Hitchens.

LAST WORD

This intrepid warrior for truth, this cultured, courteous citizen of the world, this devastating, coruscating enemy of lies and cant – well, maybe he has no immortal soul – none of us has. But in the only meaning of the words that makes any sense, the soul of Christopher Hitchens is among the immortals.

Sources and acknowledgements

The author, editor and publishers gratefully acknowledge the permission of copyright holders to reproduce material in this volume.

I. The value(s) of science

'The values of science and the science of values': Edited version of Amnesty Lecture delivered at the Sheldonian Theatre, Oxford, on 30 January 1997 and subsequently published as ch. 2 of Wes Williams, ed., *The Values of Science: Oxford Amnesty Lectures 1997* (Boulder, Colo., Westview Press, 1998). Reproduced with permission of Westview Press.

'Speaking up for science: an open letter to Prince Charles': Originally published in John Brockman's online salon *The Edge*, www.edge.org, and in *The Observer*, 21 May 2000.

'Science and sensibility': Originally delivered as a lecture at the Queen Elizabeth Hall, London, 24 March 1998, and broadcast on BBC Radio 3 as part of the series 'Sounding the Century: what will the twentieth century leave to its heirs?'

'Dolittle and Darwin': Abbreviated version of text first published in John Brockman, ed., *When We Were Kids: how a child becomes a scientist* (London, Cape, 2004).

II. All its merciless glory

'"More Darwinian than Darwin": the Darwin–Wallace papers': Slightly abridged version of speech given on 26 November 2001 at the Royal Academy of Arts, London, and published in *The Linnean*, vol. 18, 2002, pp. 17–24.

'Universal Darwinism': Slightly abridged version of speech at the 1982 Darwin Centenary Conference in Cambridge, subsequently published as a chapter under the same title in D. S. Bendall, ed., *Evolution from Molecules to Men* (Cambridge, Cambridge University Press, 1986). Reproduced with permission.

'An ecology of replicators': Slightly abbreviated text of an essay first published in special issue of *Ludus Vitalis* celebrating the centenary of Ernst Mayr: Francisco J. Ayala, ed., *Ludus Vitalis: Journal of Philosophy of Life Sciences*, vol. 12, no. 21, 2004, pp. 43–52.

'Twelve misunderstandings of kin selection': Abbreviated version of paper first published in *Zeitschrift für Tierpsychologie: Journal of Comparative Ethology*, vol. 51, 1979, pp. 184–200 (Verlag Paul Parey, Berlin und Hamburg).

III. Future conditional

'Net gain': First published in John Brockman, ed., *Is the Internet Changing the Way You Think? The net's impact on our minds and future*, Edge Question series (New York, Harper Perennial, 2011).

'Intelligent aliens': First published in John Brockman, ed., *Intelligent Thought: science versus the Intelligent Design movement* (New York, Vintage, 2006), pp. 92–106.

'Searching under the lamp-post': First published on the website of the Richard Dawkins Foundation for Reason and Science, 26 December 2011.

'Fifty years on: killing the soul?': First published as 'The future of the soul', in Mike Wallace, ed., *The Way We Will Be Fifty Years from Today* (Nashville, Tenn., Thomas Nelson, 2008), pp. 206–10. Copyright © 2008 by Mike Wallace and Bill Adler. Used by permission of Thomas Nelson. www.thomasnelson.com.

IV. Mind control, mischief and muddle

'The "Alabama Insert"': First published in *Journal of the Alabama Academy of Science*, vol. 68, no. 1, 1997, pp. 1–19. A revised version was published as 'The "Alabama Insert" by Richard Dawkins', excerpted from *Charles Darwin: a celebration of his life and legacy*, edited by James Bradley and Jay Lamar (Montgomery, Ala., NewSouth Books, 2013).

'The guided missiles of 9/11': First published in the *Guardian*, 15 September 2001.

'The theology of the tsunami'. First published in *Free Inquiry*, April/May 2005.

'Merry Christmas, Prime Minister!': First published as 'Do you get it now, Prime Minister?', in *New Statesman*, 19 December 2011–1 January 2012.

'The science of religion': Abbreviated text of the first of two lectures given at Harvard University in 2003 in the series of 'Tanner Lectures on Human Values' and published in G. B. Peterson, ed., *The Tanner Lectures on Human Values* (Salt Lake City, University of Utah Press, 2005).

'Is science a religion?': Edited text of a speech given to the American Humanist Association in Atlanta, Georgia, in 1996, in acceptance of their award of Humanist of the Year, and published in *The Humanist*, 1 Jan. 1997.

'Atheists for Jesus': First published in *Free Inquiry*, December 2004–January 2005.

V. Living in the real world

'The dead hand of Plato': This article is largely taken from 'The tyranny of the discontinuous mind', in *New Statesman*, Christmas double issue, 2011, combined with parts of 'Essentialism', in John Brockman, ed., *This Idea Must Die: scientific theories that are blocking progress*, Edge Question series (New York, HarperCollins, 2015).

'"Beyond reasonable doubt"?': First published as 'O J Simpson wouldn't be so lucky again', in *New Statesman*, 23 January 2012.

'But can they suffer?': First published on boingboing.net, 30 June 2011.

'I love fireworks, but . . .': A version of this article was published in the *Daily Mail*, 4 November 2014.

'Who would rally against reason?': Originally published in *Washington Post*, 21 March 2012; reproduced with minimal changes on the website of the Richard Dawkins Foundation for Reason and Science on 31 May 2016 (https://richarddawkins.net/2016/05/who-would-rally-against-reason/).

'In praise of subtitles; or, a drubbing for dubbing': A slightly abridged version was first published in *Prospect*, August 2016.

'If I ruled the world . . .': First published in *Prospect*, March 2011.

VI. The sacred truth of nature

'About time': Text of speech given to open an exhibition of the same title held by the Ashmolean Museum in Oxford in 2001, and published in the *Oxford Magazine*, 2001.

'The giant tortoise's tale: islands within islands': First published in the *Guardian*, 19 February 2005.

'The sea turtle's tale: there and back again (and again?)': First published in the *Guardian*, 26 February 2005.

'Farewell to a digerati dreamer': First published as foreword to Douglas Adams and Mark Carwardine, *Last Chance to See*, new edn (London, Arrow, 2009).

VII. Laughing at live dragons

'Fundraising for faith': First published in *New Statesman*, 2 April 2009.

'The Great Bus Mystery': First published in Ariane Sherine, ed., *The Atheist's Guide to Christmas* (London, HarperCollins, 2009). Reprinted by permission of HarperCollins Publishers Ltd. © author 2009.

'Jarvis and the Family Tree': written in 2010; previously unpublished.

'Gerin Oil': First published in *Free Inquiry*, December 2003, and then abridged, as 'Opiate of the masses', in *Prospect*, October 2005.

'Sage elder statesman of the dinosaur fancy': First published as foreword to Robert Mash, *How to Keep Dinosaurs*, 2nd edn (London, Weidenfeld & Nicolson, 2003). Reproduced with permission of The Orion Publishing Group, London. Foreword © Richard Dawkins, 2003.

'Athorism: let's hope it's a lasting vogue': First published in *Washington Post*, 1 January 2007.

'Dawkins' Laws': Response to 2004 the Edge annual question, 'What is your Law?': https://www.edge.org/annual-question/whats-your-law.

VIII. No man is an island

'Memories of a maestro': Text of opening speech at conference to honour Niko Tinbergen, 20 March 1990, subsequently published as the introduction to M. S. Dawkins, T. R. Halliday and R. Dawkins, eds, *The Tinbergen Legacy* (London, Chapman & Hall, 1991).

'O my beloved father: John Dawkins, 1915–2010': First published as 'Lives remembered: John Dawkins', *Independent*, 11 December 2010. © *The Independent*, www.independent.co.uk.

'More than my uncle, A. F. "Bill" Dawkins, 1916–2009': Eulogy delivered at St Michael and All Angels Church, Stockland, Devon, 11 November 2009.

'Honouring Hitch': Speech made at ceremony to present the Atheist Alliance of America's Richard Dawkins Award to Christopher Hitchens at the Texas Freethought Convention, 8 October 2011.

Bibliography of works cited

The following list gives publication details of works mentioned in the text and footnotes.

Adams, Douglas, *The Restaurant at the End of the Universe*
  (London, Pan, 1980)
Adams, Douglas, and Carwardine, Mark, *Last Chance to See*, new edn
  (London, Arrow, 2009)
Axelrod, Robert, *The Evolution of Cooperation*, new edn
  (London Penguin, 2006)
Barker, Dan, *God: the most unpleasant character in all fiction*
  (New York, Sterling, 2016)
Barkow, J. H., Cosmides, L., and Tooby, J., eds, *The Adapted Mind*
  (Oxford, Oxford University Press, 1992)
Cartmill, Matt, 'Oppressed by evolution', *Discover*, March 1998.
Cronin, Helena, *The Ant and the Peacock: altruism and sexual selection from
  Darwin to today* (Cambridge, Cambridge University Press, 1991)
Dawkins, Marian Stamp, *Animal Suffering* (London, Chapman & Hall, 1980)
Dawkins, Marian Stamp, *Why Animals Matter: animal consciousness, animal
  welfare, and human well-being* (Oxford, Oxford University Press, 2012)
Dawkins, Richard, *The Ancestor's Tale: a pilgrimage to the dawn of life*
  (London, Weidenfeld & Nicolson, 2004; 2nd edn with Yan Wong, 2016)
Dawkins, Richard, *An Appetite for Wonder: the making of a scientist*
  (London, Bantam, 2013)
Dawkins, Richard, *The Blind Watchmaker*
  (London, Longman, 1986)
Dawkins, Richard, *Brief Candle in the Dark: my life in science*
  (London, Bantam, 2015)
Dawkins, Richard, *Climbing Mount Improbable*
  (London, Viking, 1996)
Dawkins, Richard, *A Devil's Chaplain* (London, Weidenfeld & Nicolson, 2003)
Dawkins, Richard, *The Extended Phenotype*
  (London, Oxford University Press, 1982)

Dawkins, Richard, *The God Delusion* (London, Bantam, 2006; 10th anniversary edn, London, Black Swan, 2016)

Dawkins, Richard, *The Greatest Show on Earth: the evidence for evolution* (London, Bantam, 2009)

Dawkins, Richard, *River Out of Eden* (London, Weidenfeld & Nicolson, 1994)

Dawkins, Richard, *The Selfish Gene* (Oxford, Oxford University Press, 1976)

Dawkins, Richard, *Unweaving the Rainbow* (London, Allen Lane, 1998; pb Penguin, 1999)

Dennett, Daniel C., *Elbow Room: the varieties of free will worth wanting* (Oxford, Oxford University Press, 1984)

Dennett, Daniel C., *Freedom Evolves* (New York, Viking, 2003)

Dennett, Daniel C., *From Bacteria to Bach and Back* (London, Allen Lane, 2017)

Edwards, A. W. F., 'Human genetic diversity: Lewontin's fallacy', *BioEssays*, vol. 25, no. 8, 2003, pp. 798–801

Glover, Jonathan, *Causing Death and Saving Lives* (London, Penguin, 1977)

Glover, Jonathan, *Choosing Children: genes, disability and design* (Oxford, Oxford University Press, 2006).

Glover, Jonathan, *Humanity: a moral history of the twentieth century* (London, Cape, 1999)

Gould, Stephen J., *Full House* (New York, Harmony, 1996)

Gould, Stephen J., *Hen's Teeth and Horse's Toes* (New York, Norton, 1994)

Gross, Paul R., and Levitt, Norman, *Higher Superstition: the academic left and its quarrels with science* (Baltimore, Johns Hopkins University Press, 1994)

Haldane, J. B. S., 'A defence of beanbag genetics', *Perspectives in Biology and Medicine*, vol. 7, no. 3, Spring 1964, pp. 343–60

Harris, Sam, *The Moral Landscape: how science can determine human values* (London, Bantam, 2010)

Hitchens, Christopher, *The Missionary Position: Mother Teresa in theory and practice* (London, Verso, 1995)

Hoyle, Fred, *The Black Cloud* (London, Penguin, 2010; first publ. Heinemann, 1957)

Hughes, David P., Brodeur, Jacques, and Thomas, Frédéric, *Host Manipulation by Parasites* (Oxford, Oxford University Press, 2012)

Huxley, Julian, *Essays of a Biologist* (London, Chatto & Windus, 1926)

Huxley, T. H., and Huxley, J. S., *Touchstone for Ethics* (New York, Harper, 1947)

Kimura, Motoo, *The Neutral Theory of Molecular Evolution*
(Cambridge, Cambridge University Press, 1983)
Langton, C., ed., *Artificial Life* (Reading, Mass., Addison-Wesley, 1989)
Mayr, Ernst, *Animal Species and Evolution*
(Cambridge, Mass., Harvard University Press, 1963)
Mayr, Ernst, *The Growth of Biological Thought: diversity, evolution and inheritance* (Cambridge, Mass., Harvard University Press, 1982)
Orians, G., and Heerwagen, J. H., 'Evolved responses to landscapes', in Barkow et al., eds, *The Adapted Mind*, ch. 15.
Pinker, Steven, *The Better Angels of our Nature: why violence has declined* (London, Viking, 2009; pb, subtitled *A history of violence and humanity*, London, Penguin, 2012)
Pinker, Steven, *How the Mind Works* (London, Allen Lane, 1998)
Pinker, Steven, *The Language Instinct* (London, Viking, 1994)
Rees, Martin, *Before the Beginning* (London, Simon & Schuster, 1997)
Ridley, Mark, *Mendel's Demon: gene justice and the complexity of life* (London, Weidenfeld & Nicolson, 2000; published in US as *The Cooperative Gene*, New York, Free Press, 2001)
Ridley, Matt, *The Origins of Virtue: human instincts and the evolution of cooperation* (London, Penguin, 1996)
Rose, S., Kamin, L. J., and Lewontin, R. C., *Not in our Genes* (London, Penguin, 1984)
Sagan, Carl, *The Demon-Haunted World* (London, Headline, 1996)
Sagan, Carl, *Pale Blue Dot* (New York, Ballantine, 1996)
Sahlins, Marshall, *The Use and Abuse of Biology: an anthropological critique of sociobiology* (Ann Arbor, Mich., University of Michigan Press, 1977)
Shermer, Michael, *The Moral Arc: how science and reason lead humanity toward truth, justice and freedom* (New York, Holt, 2015)
Singer, Charles, *A Short History of Biology* (Oxford, Oxford University Press, 1931)
Wallace, Alfred Russel, *The Wonderful Century: its successes and failures* (New Jersey, Dodd, Mead & Co, 1898)
Washburn, S. L. 'Human behavior and the behavior of other animals', *American Psychologist*, vol. 33, 1978, pp. 405–18
Weinberg, Steven, *Dreams of a Final Theory: the search for the fundamental laws of nature* (London, Hutchinson, 1993)
Weiner, Jonathan, *The Beak of the Finch: a story of evolution in our time* (pb New York, Vintage, 2000)
Wells, H. G., *Anticipations of the Reaction of Mechanical and Scientific Progress upon Human Life and Thought* (London, Chapman & Hall, 1902)

Williams, George, *Adaptation and Natural Selection: a critique of some current evolutionary thought* (Princeton, 1966)

Williams, George C., *Natural Selection: domains, levels and challenges* (Oxford, Oxford University Press, 1992)

Wilson, Edward O., *On Human Nature* (Cambridge, Mass., Harvard University Press, 1978)

Wilson, Edward O., *The Social Conquest of Earth* (New York, Liveright, 2012)

Wilson, Edward O., *Sociobiology* (Cambridge, Mass., Harvard University Press, 1975)

Winston, Robert, *The Story of God: a personal journey into the world of science and religion* (London, Bantam, 2005)

Index

Richard Dawkins is an ethologist and evolutionary biologist and from 1995 to 2008 was the University of Oxford's Professor for Public Understanding of Science. He was first catapulted to fame with his iconic work of 1976, *The Selfish Gene*, which he followed with a string of prestigious bestselling books including *The Blind Watchmaker*, *Climbing Mount Improbable*, *The Ancestor's Tale* and *The God Delusion*. He is also the author of the anthology *A Devil's Chaplain* and two volumes of autobiography, *An Appetite for Wonder* and *Brief Candle in the Dark*. He is a Fellow of both the Royal Society and the Royal Society of Literature, and the recipient of numerous honours and awards. He remains a fellow of New College, Oxford. In 2013, Dawkins was voted the world's top thinker in *Prospect* magazine's poll of 10,000 readers from over a hundred countries.

THE GOD DELUSION
Richard Dawkins

A timely, impassioned and brilliantly argued polemic on atheism.

'A very important book, especially in these times . . . a
magnificent book, lucid and wise, truly magisterial'
Ian McEwan

'An entertaining, wildly informative, splendidly written
polemic . . . we are elegantly cajoled, cleverly harangued into
shedding ourselves of this superstitious nonsense that has
bedevilled us since our first visit to Sunday school'
Rod Liddle, *Sunday Times*

'A spirited and exhilarating read . . . Dawkins comes roaring
forth in the full vigour of his powerful arguments'
Joan Bakewell, *Guardian*

'Passionate, clever, funny, uplifting and
above all, desperately needed'
Daily Express

'A wonderful book . . . joyous, elegant, fair, engaging, and often
very funny . . . informed throughout by an exhilarating
breadth of reference and clarity of thought'
Michael Frayn

'Everyone should read it. Atheists will love Mr Dawkins's
incisive logic and rapier wit'
Economist

'Richard Dawkins's *The God Delusion* should be read by everyone
from atheist to monk. If its merciless rationalism doesn't
enrage you at some point, you probably aren't alive'
Julian Barnes

'There is not a dull page in Richard Dawkins's *The God
Delusion*, a book that makes me want to cheer its
clarity, intelligence and truth-telling'
Claire Tomalin